Mapping Geomorphological Environments

Kosmas Pavlopoulos · Niki Evelpidou ·
Andreas Vassilopoulos

Mapping Geomorphological
Environments

 Springer

Dr. Kosmas Pavlopoulos
Harokopion University
    of Athens
Dept. Geography
El. Venizelou Str. 70
176 71 Athens
Greece
kpavlop@hua.gr

Dr. Andreas
Vassilopoulos
GeoEnvironmental Institute
Flias & V. Ipirou Str. 13
151 25 Athens
Greece
vassilopoulos@gcparks.com

Dr. Niki Evelpidou
University of Athens
Fac. Geology &
    Geoenvironment
Dept. Geography & Climatology
157 84 Zografou
Panepistimiopolis
Greece
evelpidou@geol.uoa.gr

ISBN 978-3-642-01949-4   ISBN 978-3-642-01950-0 (eBook)
DOI 10.1007/978-3-642-01950-0
Springer Dordrecht Heidelberg London New York

Library of Congress Control Number: 2009927029

Cover design: deblik, Berlin

Printed on acid-free paper

Springer is part of Springer Science+Business Media (www.springer.com)

Design by

MSc Chartidou Konstantia
Archaeologist - GIS expert
University of Franche Compte
Institute des Sciences et
Techniques de l'Antiquite

Efraimiadou Eleni
Geologist
University of Athens
Faculty of Geology &
Geoenvironment

Saridakis Nikolaos
Graphics Designer
University of Athens
Faculty of Geology &
Geoenvironment

MSc. Matiatos Ioannis
Geologist
University of Athens
Faculty of Geology &
Geoenvironment

MSc. Koutsomichou Ioanna
Geologist
University of Athens
Faculty of Geology &
Geoenvironment

Katsiadramis Stamatis
Geographer
Harokopio University of Athens
Faculty of Geography

Giotitsas Ilias
Geologist
University of Athens
Faculty of Geology &
Geoenvironment

Nika Konstantina
Geologist
University of Athens
Faculty of Geology &
Geoenvironment

Skentos Athanassios
Geographer
Harokopio University of Athens
Faculty of Geography

Konstantopoulou Zeta
Graphics Designer
Technological Institute of Athens
Faculty of Graphic Design

The project is co-funded by the European community (75%)

Education and Culture

Leonardo da Vinci

The book doesn't necessarily reflect the views of the European Commission or the National Agency

Table of Contents

FOREWORD   ix

introduction   1

methodology-techniques   5

ADVANCES IN GEOMORPHOLOGICAL MAPPING   6

The basics of geomorphological mapping   6

Mapping from 1900 to 2000   7

Geomorphological mapping in the 21st century   12

The value of a geomorphological map in applied geology   44

fluvial environments   49

FLUVIAL PROCESSES   50

Rivers, water streams and fluvial processes   50

Base level   54

Geomorphological development of valley systems   54

Transfered material deposits by waterstreams   55

Fluvial geomorphological cycle   57

Erosion cycle   58

Terraces   59

Disruption of the erosion cycle – Rejuvenation stage   60

MAIN FLUVIAL LANDFORMS   62

coastal environments   69

COASTAL PROCESSES   70

Sea water as factor of the coastline formation   70

Retreat - Coastal Erosion   70

Waves   71

Coastal currents   72

Sources of coastal sediments - Balance of the coastal zone sediments 74

Coastal sediments balance 75

Sea level changes 76

Classification of coasts 78

Coastal lagoon systems 82

Evolution of estuary systems 84

Internal Circulation - Hydrodynamics 84

Coastal sediments 85

MAIN COASTAL LANDFORMS 86

lacustrine environments 99

LACUSTRINE PROCESSES 100

Lakes-Introduction 100

History of the existence of lakes 100

Classification of lakes 101

Lake water: Compostion, Movements and Properties 104

Sedimentation in lake environments 106

MAIN LACUSTRINE LANDFORMS 108

glacial environments 111

GLACIAL PROCESSES 112

Creation and expansion of glaciers 112

Glacial weathering and erosion 113

Glacial deposits 115

Periglacial areas 118

Glacial and Eustatic processes 118

Expansion of glaciers during the Quaternary 120

Modern glaciers   123

MAIN GLACIAL LANDFORMS   124

karstic environments   135

KARSTIC PROCESSES   136

Karst - Introduction   136

Dissolution of limestones   137

Karstic geomorphology   139

Forms of dissolution   141

Karstic evolution cycle   142

MAIN KARSTIC LANDFORMS   144

volcanic environments   151

VOLCANIC PROCESSES   152

Volcanism   152

Types of volcanic explosions   153

MAIN VOLCANIC LANDFORMS   158

aeolian environments   165

AEOLIAN PROCESSES   166

Aeolian transport and deposition   166

Sand and wind interaction   167

MAIN AEOLIAN LANDFORMS   170

surface landforms   175

topography, lithology and tectonics   185

TOPOGRAPHY   186

SEDIMENTARY FORMATIONS   188

METAMORPHIC ROCKS   191

TECTONICS   192

geomorphological mapping (case studies)   197

Case study 1: Geomorphological study of Oinois river
(North Attica-Greece)   198

Methodology   199

Geomorphological mapping   199

Geomorphology of the coastal alluvial fan   200

Geomorphological characteristics of the coastal zone   201

Human activities in the coastal zone   202

Case study 2: Geomorphological study of Attica basin (Greece)   203

The Athens plain   205

Case study 3: Geomorphological study of Paros Island (Greece)   206

Case study 4: Geomorphological study of Southern
Attica (Greece)   209

Geology   210

Tectonics   213

Climatic conditions   213

Relations between relief, climate and hydrography   214

Conclusions   216

REFERENCES   227

Foreword

In the last few years, Geomorphology, like the rest of the Geosciences, has developed at a enormous rate. This development was due to an interdisciplinary opening that was made towards environmental sciences, ecology, archaeology and management. In all these new disciplinary fields, the geomorphological map has become an essential tool in order to understand the environment's dynamics but also in order to help in the decision making.

Nowadays, the realisation of the geomorphological map is profiting from the modern tools of informatics, which are computer designing or computer mapping, that is, geographical information systems. Consequently, in order to conceive and realise a geomorphological map adapted to the user's needs, the possibilities but also the constraints of these new tools should be taken into account.

For the first time, a work realised by geographers and geomorphologists is entirely dedicated to this fundamental need. Kosmas Pavlopoulos, Niki Evelpidou and Andreas Vassilopoulos share with us their experience in the field of geomorphological mapping, through this pedagogic, clear, well illustrated and very readable work.

The major originality of this work is that it addresses at the same time, the geomorphologists who are eager to adopt their cartographic methods and the cartographers who are eager to better understand the forms and models that they are called upon to map. In the first part, the book offers two reading paths, a first methodological one and the second one dedicated, according to the medium, successively to fluvial, littoral, lacustrine, glacial, periglacial, karstic, volcanic and aeolian environments. In the second part the book presents a series of case studies that provide concrete answers and numerous examples to the needs of geomorphological mapping.

The Greek geomorphological school has developed considerably during the last years. Kosmas Pavlopoulos, Niki Evelpidou and Andreas Vassilopoulos give a brilliant example through this work.

Eric Fouache

President of GFG

Chairman of the Working group on Geoarchaeology «IAG»

July 2008

Niagara falls - Canada (by N. Tsoukalas)

introduction

Surface waters play an important role in relief formation by creating a multitude of landforms which depend genetically and evolutionally on the prevailing geomorphic processes and on the area's geology. Underground waters, in turn, form a series of underground landforms and deposits which depend on geomorphic processes different from those prevailing on the surface.

Water, through infiltration in geological formations, follows a course which depends on many parameters and forms what one would call a «underground relief».

Water is important in all its states, and necessary for all known forms of life on our planet. Its quality and physicochemical properties form the environment of the ecosystems of which it is component. It is characteristically mentioned that water is the most common solvent in the terrestrial system, as it dissolves and transports a wide variety of chemical substances (salts, minerals, etc). It significantly interferes in the chemical decomposition of rocks and in soil formation; it also has high heat capacity, thus influencing the environment.

The quantity of water on our planet is practically stable and amounts about $1,600 \times 10^6$ Km^3. Fresh water represents 0.6% Km^3 of this quantity (or in other words 8.2×10^6) but only 0.1×10^6 Km^3 of surface water and 3×10^6 Km^3 of underground water are available to man for use.

The water cycle is one of the most important of nature's cycles in progress. In its simplest description, it includes water evaporation of oceans, lakes, rivers, etc, transportation and condensation of water vapours within the atmosphere, its reintroduction onto the Earth's surface in the form of rain, snow or ice, and finally its surface runoff or infiltration and underground flow until it reaches the «basic» sea level.

Water reserves are not unlimited, particularly those of fresh water, which is necessary for the viability of many ecosystems, and also for human survival. Water resources management constitutes one of the most crucial ecological problems (i.e. water shortage, water pollution, etc).

Geomorphology, through mapping techniques but also through analysis and understanding of geomorphic processes, contributes to the issues of water resources management, and to issues related both to the hydrological and hydrogeological cycle. An area's geomorphological evolution is directly connected to water runoff, flood yields, estuary systems, areas under erosion, transportation and deposition. The hydrological and administrative researches of an area's drainage network and water resources are carried out based on geomorphological research and mapping.

The geomorphological analysis and mapping of karstic areas, is the basis for management planning, aiming both at the preservation of the geological and geomorphological heritage (caves, karstic forms), and the preservation and protection of underground karstic aquifers. Karst geomorphology and the understanding of karstic systems' evolution are important scientific tools for hydrogeology. In glacial and periglacial environments

Flooded plain in Hungary (by C. Centeri).

geomorphological mapping is necessary in order to understand climatic changes, and is also the basis for the development of a protection and preservation program for these environments. Coastal geomorphological mapping and research, in combination with coastal dynamics, are the basis for the creation of a continuous registration and control network in order to carry out complete and systematic coastal zone management. Such a network will supply the coastal zone evolution and management models with data. The geomorphological evolution of the Earth's surface is strongly connected to its underground evolution (caves, karst channels, wells, etc.). The landforms and deposits that were created by endogenous and exogenous evolutional processes are being reformed through time.

Geomorphology and particularly geomorphological mapping, provides the ability to identify, impress, and analyse landforms and to associate them to the evolution processes of both superficial and underground relief. The utility and necessity of geomorphological cartography in the study of superficial and underground waters and in their management becomes more and more imperative because of the increasing interference of humans with the environment.

Chapter 1

methodology-techniques

ADVANCES IN GEOMORPHOLOGICAL MAPPING

The basics of geomorphological mapping

Geomorphology presents great complexity because of the numerous approaches to geomorphological analysis and the wide variety of geomorphological mapping scales. The nature of the geomorphic unit is controlled both by the chosen analysis model and the mapping scale required. The two main features fundamental for the basic geomorphic unit are homogeneity and indivisibility at the chosen scale. The basic geomorphic unit should have homogeneity, and may be defined in terms of genetic or structural pattern, which is the approach followed by the IGU (International Geographical Union) and most European geomorphologists. There is also an alternative used by British system followers, according to which the location and dimensions of geometric elements play an important part.

Most detailed geomorphological maps are developed for small areas on quite large scales, typically between 1:10.000 and 1:50.000. Regional analysis of landforms is a very significant aspect of modern geomorphology, and implies large scale regionalisation of geomorphological maps.

There are two Models, the Landform Elements Model and the Landform Patterns Model, which are complementary approaches to the analysis of geomorphic units. Every land section can be described by both models, with the choice depending upon the scale and purpose of the mapping.

According to the Landform Elements Model, the units of landscape are compared to "a simply curved geometric surface without inflections" focused on slope and slope measurements. According to the Landform Patterns Model, the land surface is seen as a "3D cyclic or repetitive phenomenon" in which simpler elements recur at quasi-regular intervals in a definable pattern and the elements that form the patterns are identified as units. The landform elements model used by Greek and British geomorphologists classifies the basic units of landscape in geometric terms as facets and segments defined by slope and area measurements. Systems similar to those models, with slight variations have been developed by national groups.

Differences in the identification of geomorphic units are significantly related to matters of regionalisation and scale. The identification of different features as basic homogeneous units is the product of different scale use in order to cover regions of different size. The choice of unit depends on the scale of analysis. The clearest and simpliest classification basis is the "classification of landscapes into homogeneous units suitable to the mapping scale required for the particular purpose". In most regions, a hierarchy of land units can be identified, depending on the mapping scale. The landscape could be considered as a multi-tiered geosystem, where each tier consists of different taxonomic individuals that form the basic geomorphic units of the landscape. By using smaller scales in the study of wider regions, smaller features and processes often fade from view while larger features, imperceptible at larger scales, become apparent.

Mapping from 1900 to 2000

The study of landforms, their structure and development, includes the need to illustrate both the findings of an investigation and the character of the landforms investigated. A wide variety of illustration methods that includes sketches, block diagrams, and various types of photography and other imagery, both from the ground and from the air, has been used by geomorphologists, in order to describe the Earth's land surface. Attempts made recently by many geomorphologists to develop a graphical display method for the Earth's physical surface features have finally led to the creation of various geomorphological map forms. Many European practitioners contributed to this creation. These detailed maps are more than a means of illustration; they are a major research instrument in both theoretical and applied geomorphology.

Throughout the 19th century and into the early years of the 20th century, the principal method for studying landforms was through static descriptive physiography. Some researchers, in Europe and the United States, recognised the influence of dynamic forces on landscape (e.g. John Wesley Powell saw the force of water in the erosion process of the Grand Canyon).

In 1899, William Morris Davis published "The Geographical Cycle" where, for the first time, was stated the basic concept of the "cycle of erosion", which produced a radical change in geomorphology. He introduced the concept that landscape was dynamic and constantly evolving in a cycle due to external forces. Davis' dynamic approach to landscape left his mark on geomorphology, although his initial theory was not originally adopted by the scientific community but was instead negatively criticised. However, in practice, static descriptive physiography was still the primary way of carrying out most geomorphological research; with landscape being described in writing, generally accompanied by artistic block diagrams drawn to illustrate the author's conclusions. Although these diagrams were often excellent illustrations of geomorphological processes, they tended to be qualitative designs rather than quantitative verifiable graphic analyses of landscape.

Photographic quality and analysis reached a high level of sophistication by the 1920s and thus started being useful to geomorphologists. Until the early 1840s, when photography was recognised as a powerful tool in topographic mapping, there were only a few photography users in landform study. Aerial photography, although, experimental, was first introduced as a means of landscape study, by the early photographers and balloonists, Nadar and Triboulet.

Albert Heim was the first to use aerial photography in geomorphological research and in 1899, he published his photographs and observations, which were made during a balloon flight over the Alps. Aerial photographs were widely used during the First World War for giving a view of the enemy's area and spotting battlefield positions.

In the early 20th century, before World War II, specific sites or factors of landscape were the target of most geomorphological research. The land was not examined from a broader point of view. Gradually, researchers started to focus on the visualisation of regional landscapes by means of physiographic or landform maps. These combined elements of maps topographic and geomorphic units. In the form of block diagram, the physiographic map depicts actual landforms in perspective, from an oblique view point.

World War II marked a breakthrough in geomorphology, which diversifed both theoretically and technically. The very dynamic warfare style of World War II which implied rapidly moving war units, fast thinking and decision making required very detailed analysis of the enemy's terrain, which lead to great technological advances in use and interpretation of aerial photography. The ability to study landforms and analyse them, by means of aerial photographs increased with the improvement of photo equipment, films and interpretation instruments.

In 1957, quantitative analytical techniques, which were initially developed for other scientific purposes, were applied into geomorphologic research. Geomorphologists, particularly in Europe, became interested in wide-ranging analysis of landforms that considered all aspects and features of landscape together.

It became apparent that landform development processes were more complex than the relatively simple Davisian cycle of uplift, downwearing to a peneplain, and rejuvenation.

Attempts were made to relate past landform processes to present ones. How could various landscapes be compared and contrasted to one another? What was the effect of landform and relief, on vegetation, hydrology, and the cultural development of the area? These among other issues brought to light the need for a new paradigm in geomorphology. The complexity of the landscape led geomorphologists to attempt the establishment of an objective scientific method for the graphic portrayal of complex landform factors, in order to pursue orderly scientific research. The necessary information for detailed and accurate analytical study of landforms was not provided by the qualitative descriptions and artistic diagrams of the physiographers.

In the 1950s and 1960s, the science of geomorphology developed into analytic physiography of the Earth's surface and the detailed geomorphological map became the main research tool in geomorphology.

The main aspects of research in modern analytic geomorphology are five fundamental landform concepts:

- *Morphography (or Morphology):* the appearance, shape, etc. of the landscape. This refers to the qualitative description and the geometric elements of landforms. It is the principal feature of the descriptive geomorphological analysis and should be carried out with the maximum possible precision. Information should be provided wherever possible that quantifies the landforms.
- *Morphometry:* measurements,

dimensions, and slope values of landforms. This mainly refers the the quantitative elements : altitude, relief and slope inclinations, landform borders, angles and lengths of linear cartographic elements (tectonic discontinuities, branches of drainage networks, etc), surface covered by planation surfaces, karstic and volcanic landforms, etc.

- *Morphogenesis:* the origin of each landform. This concerns the genetic processes, morphogenetic systems and mathematical simulations that form an area's relief over time.

- *Morphochronology:* the age of each landform. Absolute and relative dating, sediments' correlation, landforms' grouping and correlation based on their age and position.

- *Morphodynamics:* the land-forming processes currently active or those that may be activated in the future. This refers to all the dynamic processes which form the earth's relief. They are usually identified as "traces", remains of past dynamic processes (inherited landforms).

The graphic portrayal of these five concepts involves a complex and often difficult set of analytical and cartographic procedures. The development of theory, procedures, and cartographic legends has taken a great deal of time and effort, particularly by European geomorphologists, over the last 30 years. There is an obvious need for detailed geomorphological maps in order to produce further geomorphological research and enhance the value of geomorphology in applied landscape analysis.

However, there are many views about the correct character and content of geomorphological maps, quite different from one another.

Some of the earliest detailed geomorphological maps were published in 1914 in Siegfried Passarge's Morphological Atlas. Since then and up until the end of World War II a gap occurred, very few detailed local maps were published by European geomorphologists and detailed geomorphological surveys were only occasionally made. Not until after the 18th Congress of IGU (International Geographical Union) in 1956, did the importance of detailed geomorphological maps received international acknowledgment. Two years later, three tasks were given to the newly created Subcommission on Geomorphological Mapping at the IGU congress in Stockholm:

- The introduction and development of the methodology of geomorphological mapping.

- The international adoption of a uniform system for geomorphological mapping in order to ensure compatibility.

- The demonstration of geomorphological mapping applications in regional economic planning, for the facilitation of a rational utilisation of the Earth's surface.

A large number of countries, including Switzerland, U.S.S.R., Poland, France, Czechoslovakia, Japan, Belgium, and Hungary were preparing detailed geomorphological maps, before the subcommission was formed. The content and methodology was different in different countries, so the maps were not generally comparable, therefore their use for geomorphological analysis of wider areas was limited and inadequate. European geomorphologists recognised the need for a single unified technique, including a common legend, for comprehensive mapping. A subcommission meeting was held in Krakow, Poland, in 1962, where representatives from 15 countries established a set of guidelines for geomorphological map preparation. These guidelines included:

- Field work as a basic necessity with aerial photographs as a recommended tool.
- Mapping at scales between 1:10,000 and 1:100,000; at these scales "relief and its peculiarities can be represented".
- Mapping of all relief aspects: morphography, morphometry, morphogenesis, and morphochronology, in order to study relief's past, present, and future development.
- The use of both colour and symbols to convey information.
- The establishment of chronological order in landforms development.
- The inclusion of lithological data.
- The arrangement of the map legend in a genetic-chronological order.
- The recognition that detailed geomorphological maps are an indispensable tool for the future development of geomorphology.

The Subcommission on Geomorphological Mapping met regularly through the 1960s. In 1968, at the IGU Congress in New Delhi, India, it was upgraded to the Commission of Geomorphic Survey and Mapping. One of its main tasks would be the development of a Manual of Detailed Geomorphological Mapping and the devising of the legend for an International Geomorphological Map of Europe at a scale of 1:2.500.000. The latter was produced with the collaboration of many European geomorphologists, and was published in 1971. In 1972, "The Manual" was published. It was a compilation of articles by 20 geomorphologists.

Although collaborative work was being carried out by the Commission on Geomorphological Survey and Mapping, there was still a great deal of diversity and disagreement on the nature of geomorphological maps and their contents. The number of legends, representing different approaches and methodologies, has proliferated. Most of them represent a specific national or regional outlook and very few, if any, meet all the requirements of comprehensive geomorphological mapping. Since 1970, the theoretical and practical importance of geomorphological mapping has increased significantly due to new developments in geomorphology. The new theories of global tectonics and the development of terrestrial investigation from space have ameliorated the understanding of the competing interactions of endogenetic and exogenetic forces.

A new emphasis on "ecological geomorphology" has related the understanding of landscape relief to human life and activity. New research methods were introduced, particularly in relation to aerial photography, satellite imagery, and radar imagery; computer science has advanced in the areas of data analysis and automated cartographic

10

display. Although progress has been made, serious problems remain. Besides problems of content and methodology, the problem of appropriate scales persists, particularly in the case of small-scale large-area; here no solution has been given yet.

Geomorphological maps are divided according to Wright into two general categories: "Landform Surveys" and "Special Feature Investigations." The first category consists of various types of detailed geomorphological maps. The second category includes specialised maps such as the "Terrain Analogues" of the U.S. Army, which provide land analysis with sole regard to its trafficability with military vehicles, or Hammond's (1954) "Small Scale Continental Landform Maps" which, although their purpose is regional and their scale small, are focused on "actual characteristics of the existing surface rather than genetic interpretations". The work of Hammond can not be classified in the category of landform surveys because of its restricted view.

In different countries, the development of geomorphological mapping has followed different, paths, partially due to the different interests and emphases of geomorphologists and partially due to real or perceived differences in the landforms found in various regional settings. The lithological-structural unit was selected by some European geomorphologists, including French, Czechoslovakian, and Hungarian scientists, as the basic element in landform analysis. However some German,

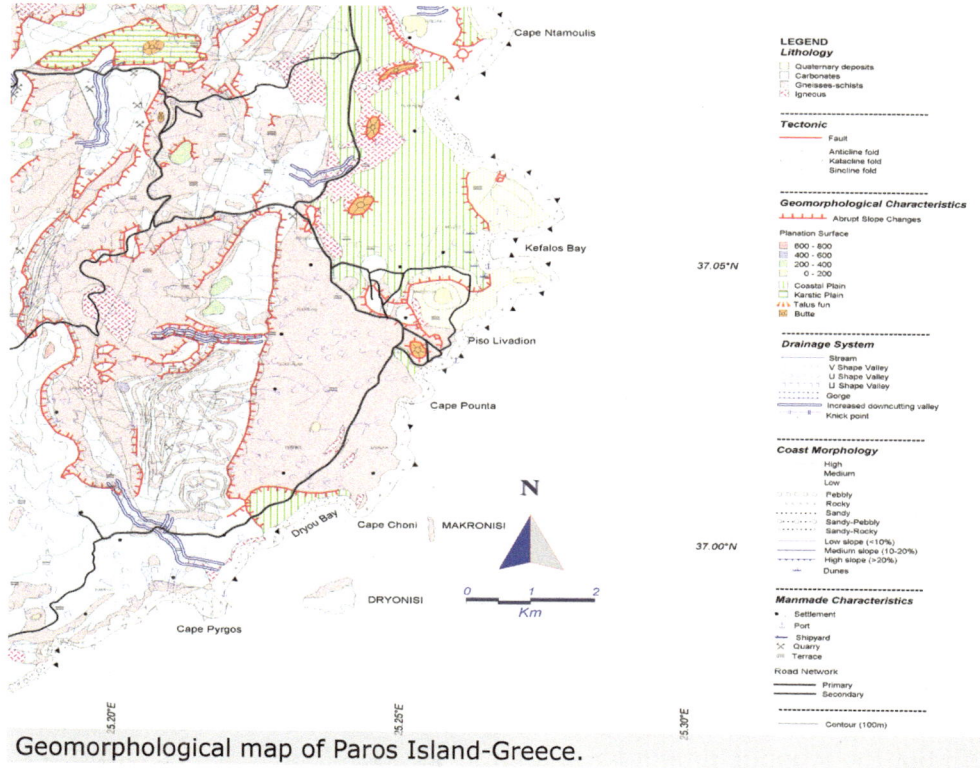

Geomorphological map of Paros Island-Greece.

Polish, Russian and Rumanian geomorphologists consider the form to be the basic unit. The Manual of Detailed Geomorphological Mapping and the legend of the International Map of Europe at 1:2.500.000 scale generally follows the Polish, Russian, and German convention. An "empirical system" was developed by geomorphologists in Great Britain, based on the division of landscape into "slopes and flats". Belgian and Canadian geomorphologists have adopted similar systems, because they can provide quantitative values to all landforms, generally at the expense of genesis and chronology. Some combinatory approaches were followed by some geomorphologists. A system of geomorphological mapping for resource survey, based on a concept of land units and land systems was developed by the Commonwealth Scientific and Industrial Research Organization (CSIRO) in Australia. The scope of the concept was to "provide both a basic and functional landscape subdivision". The term land unit represented a land surface that had similar genesis and now has similar topography, soil, vegetation, and climate . The term land system represented "an assembly of geographically and genetically related land units".

The development of the "Unified Key", an international legend and its application for the International geomorphological map of Europe, is a step forward. Although there is no universal acceptance, the Unified Key is widely accepted as the legend to use. In 1982, the desirability of a single legend for geomorphological maps on various scales was pointed out, but the legends proposed so far did not meet all the requirements. There is an underlying unity to terrain classification, although there are still many differences. The identification of the simplest land unit with low variability, and hence to combine the land units into a hierarchy of increasingly complex regions, concerns all geomorphologists.

Geomorphological mapping in the 21st century

Identifying what to map

Even with all technological advances at our disposal, geomorphological mapping still begins with the identification of the fundamental units that compose the landscape. Establishment of the nature and character of the units is critical for the success of any geomorphological research. However, there is no single agreed-upon unit that meets research mapping needs of all types and scales. A geomorphic unit in general terms is decribed as an individual, genetically homogeneous landform produced by a definite constructional or destructive geomorphic process. Most geomorphologists would agree with this definition, but would differ widely as to the descriptive characteristics

Identifying steep slopes and debris cones in a landscape.

of the genetically homogeneous landform. In addition, although most landforms may be considered as genetically homogeneous in terms of present processes, they owe their characteristics, in part to past processes of a different sorts. Each part of the land surface is the end product of an evolution governed by parent geological material, geomorphological processes, past and present climate, and time. Most geomorphologists would agree that it is necessary to view landscape in terms of recognisable repetitive patterns.

Collecting information through field survey

<u>Field sampling</u>

A. Techniques and sampling methods in geomorphological applications

In most geomorphological applications of geosciences, the descriptive and definitive techniques and approaches are not sufficient in order to determine, describe and analyse the geomorphic processes, the palaeo-environmental conditions and the palaeo-geographical spatial evolution, as regards time.

The relatively recent sediments deposited in different geoenvironments, have "registered" a set of information that can determine their palaeo-environmental conditions of deposition, as well as the post-depositional processes.

Especially, for the analysis and study of Holocene and Pleistocene depositions like talus cone, colluvial and alluvial deposits (cones, fans), fluvial-torrential, floodplain and overbank deposits, the bed and terraces deposit, the delta, lagoon and coastal deposits, as well as human deposits (archaeological layers), the sampling of loose, or even cohesive sediments, with their corresponding stratigraphical description and interpretation, is necessary.

In order to understand and analyse the sediments' chronostratigraphical sequence and depositional palaeo-environment of sediments, we must implement sampling techniques and a number of sediment analyses. Depending on the speculation and orientation of the research, they refer to mineralogical composition, grains cement, micro- and macro-fauna, microflora, dating, concentation in organic carbon, microchemistry and micromorphological features of the sediments.

B. Methodology and materials

The sampling methods can be separated into two categories according to the equipment that is being used:

• Using tools hand tools, such as, the geological hammer, chisel, knife, spatula, grater, etc.

• Using mechanical devices, such as drilling machines, vibracorring samplers, gravity samplers, etc.

In both cases, the purpose is to collect a compact sample of rock or soil, with the least possible disturbance, so that the conditions will be similar to those that prevail in the sampling location. The way of sampling depends on its purpose, which can be the determination of the sample's structure, physical, biological, chemical and mechanical features, etc. The analyses and determinations take place in specialised and certified laboratories

Sampling procedure by Vibracoring drill machine.

with scientifically acceptable methods. Usually, after sampling, a range of laboratory analyses follows, in order to determine the sample's structure, geometrical features, mineralogical composition, stratigraphical features, micro- and macro flora and fauna features, and also, a range of properties that have been "recorded" from the sample.

The sample types are distinguished in:

- *Disturbed samples:* samples that during their collection have been subjected to disturbance of soil tissue (material structure). These are suitable only for the determination of their physical properties.

- *Undisturbed samples:* samples whose collection techniques, ensure minimal soil tissue disturbance, in order to be suitable for the determination of their physical and mechanical properties. Non

disturbed samples are taken in soft cohesive soils. Shelby or Denison type soil samplers or piston soil samplers are recommended. For undisturbed samples collection:

1. Thin-walled samplers must be used, in order to minimise friction between sampler and soil

2. Equipment must be conserved and controlled diligently

3. The drill tip must be cleaned very well, with constant circulation of drilling pulp or fluid

4. The sample collector's movement must be slow, with pressure and shocks avoided

5. Sample collector's penetration must be less than its length, in order to avoid sample compression

The criterion for satisfactory or non satisfactory core sampling is the sample's collection percentage, that is, the proportion of the received sample length to the drilled length (%).

During rock drilling, the material that goes into the sample collector is divided in:

- Cores longer than 10cm
- Cores shorter than 10cm
- Rock fragments
- Material that is lost, during sample receiver lifting.

Total recovery is defined as the total length of categories the first three and is expressed as sampling length percentage.

When a large part of the core sample consists of fragments, they may be consolidated in a mass with the same core diameter and their total length measured.

Core samples after drilling.

Solid Core Recovery is called the total length of categories 1 and 2 and is expressed as sampling length percentage.

Rock Quality Designation, RQD (%) = {total core length> 10cm / sampling length} x 100.

There are various measures that can be taken during sampling, in order to decrease core loss, when the latter is due to one of the following reasons:

• Drilling post vibrations. This can be avoided by preserving the drill in good mechanical state, by decreasing the spindle's propulsion and rotation velocities and by using drilling rods of the same diameter along the full length of the drilling column.

• Excessive drilling velocity. This can be avoided by decreasing drilling and rotation velocities.

• Sample destruction because of large water circulation. This can be avoided by implementing "dry" boring in selected depths, by changing the circulating drilling pulp or fluid and by using compressed air instead of water.

• Sample pulverisation. This can

be avoided by increasing water supply, by slightly elevating the drilling column or/and lifting it to the surface to unblock it.

The term rock fragment sampling describes the systematic collection of rock decay products that are lifted to the surface with the drilling pulp or fluid that circulates while drilling. With rock fragment sampling:

• The drilling diameter remains stable independently of the rock's hardness

• There is no drilling wall collapse

• Transport of rock fragments from drilling bottom is complete with no water loss.

Rock fragments are collected on the surface, washed with water, dried, packed and sent to the laboratory for further analysis

C. Sampling with physical support

It refers to sampling in natural sections of soil materials and loose depositions by hand. It also refers to sampling with drills moved by hand (Auger type).

The first task concerns *detection and accurate determination of sampling location*, with GPS support, in order for location redetermination and sampling repetition to be possible, if required. The accurate determination of the location and also its features are necessary elements for the development of scientific research and references by other researchers.

Section cleaning follows, using tools, such as the geological hammer, spatula, grater, etc.

Then, the *stratigraphy description and planning*, of the sampling location is made, where the stratigraphic horizons and their macroscopic features (thickness, colour, composition, materials etc.) are described with the best possible accuracy. The depth from soil surface, from which sampling was made is also described.

Sampling from a specific location is the next stage. A plastic bag or a box (metal or plastic) is used, depending on whether the sample is sensitive or reacts to the conservation material and on the analysis or test to which it's going to be subjected. During this stage, if the target is an oriented and non disturbed sample, a technique using plaster bandage and perimetrical excavation should be followed.

Sample registration follows. It includes features, general information and section's photographs. It also includes the macroscopic description of the formations and its first validation.

The last stage is the sample's transport and conservation in proper conditions under which the sample's components can be kept unchangeable for future analyses.

D. Sampling with mechanical support

This way of sampling refers to the use of mechanical arrangements for sample extraction. These are divided in three categories.

Gravity devices: This usually refers to undersea samplers that are released from oceanographic vessels and are "nailed" to the buttom by gravity.

Excavation machinery: This refers to bulldozers or excavation machinery that can open trenches in loose or medium cohesive geological formations. These can be used in sections of 4- 8m deep and 1- 6m wide. Penetration to 4-8 m depths presumes artificial terracing, to create artificial slopes of smaller height, so as to achieve greater security. The construction of artificial terraces is recommended whenever the artificial slope front exceeds 3m. The life of the trench depends on quality and geotechnical features of the formation, the climatic conditions and the artificial slope charging. It can vary from several hours to several weeks. This way of sampling presumes strict security rules for researchers (helmets, large trench width, one person in the trench and two outside, ladder use, etc.).

Drill: Sampling is made with different types of drill.

1. *Flight augering* used in loose formations. With this method, soil penetration of a curved pipe with external flight spiral is achieved. The external drilling diameter is usually 75- 125mm and the penetration depth can reach up to 30- 50m. Soil samples that are collected with this method, cannot be grainy or hard, and are totally disturbed.

2. *Shock drilling, (shell and auger)*, during which, penetration into the soil (cohesive or grainy) is done with hitting shocks. In cohesive soil formations, collection of non disturbed samples is possible. In rocky formations drilling penetration is done by crushing the rock, therefore only rock fragments are recovered.

3. *Rotary drilling*, during which, drilling is made by rotating the drilling post and using cutting

heads (compact or curved), as well as special samplers that are used in combination with curved cutting heads. With this method, sampling drilling is possible,or by rotary coring, either by non-coring rotary drilling.

4. *Vibracooring sampling drill.* In this case, drilling is made by vibration and striking of the drilling rod, using cutting heads and special samplers. With this method, sampling is possible in areas that are difficult to approach; the equipment is portable and the samples are not greatly disturbed. This sampling method is normally used in medium cohesive soil formations, for small depths that do not exceed 10- 15m and for diameters smaller than 50mm. Drilling can be telescopic and, the sampler usualy has a single steam jacket, with an internal plastic pipe where the sample is collected. This methodology is suitable for geomorphological, palaeoenvironmental, palaeogeo-graphical and environmental studies using suitable samplers.

Digital field surveying

Most geosciences data is by nature three-dimensional. Despite this, traditional paper-based mapping methodologies in which 3D real-world data are simplified and displayed in 2D are used by many field geoscientists. Advanced methods have recently been developed by petroleum geologists, using high resolution seismic survey data in order to build detailed 3D models of sub-surface geological structure. One can now analyse rock outcrops exposed on the surface thanks to the development of modern optical 3D measurement and visualization techniques, by using an approach that is conceptually comparable to that used in petroleum exploration. Laser-scanners, satellite images, aerial photographs, digital photography and digital mapping methodologies provide high accuracy and spatial resolution that enable modern geomorphologists to produce detailed geomorphological maps, both in print and digital format.

These models of real-world surface are geospatially and geometrically precise and allow the geoscientists to take a precise image of the outcrop back to the laboratory where it can be visualised, analysed and interpreted. The exact geospatial position of each virtual model is achieved by the use of Real-Time Kinematic GPS, with up to one centimetre spatial precision that allows several overlapping models to be stitched together as seamlessly as possible. Final surface representations after stitching are also analysed using 3D visualisation software which allows the direct interaction with the virtual outcrop either by using full colour auto-stereoscopic 3D screens or fully immersive stereo projection.

The application of digital mapping in combination with optical 3D measurement and 3D visualisation techniques supplies geoscientists with a new set of tools that can be applied to a wide range of geological problems and has a wide range of applications and possibilities.

Effective geo-analysis is supported by the collection of high quality data concerning geological structures. Despite this, many geoscientists still find the classic paper-based mapping methodologies attractive; in a paper-based mapping environment the 3D real-world data is simplified and displayed in two dimensions. The collection of a large data volume can be realised using terrestrial laser scanning techniques which will allow geoscientists to undertake visual analysis on a scale that was never possible before. Once the Digital Terrain Model (DTM) has been created, geoscientists can visualise, analyse and interpret the model back in the laboratory.

Three dimensional large scale measurements can be applied to a broad range of geological problems, including:

- Quantitative geo-referenced 3D models for the use of geotechnical surveys into slope stability;
- The provision of sub-seismic scale, rock structure analogues, for modelling permeability and fluid flow, in hydrocarbon reservoirs;
- As lab-based assistance for the training and teaching of students and professional geoscientists in the complex geometry of structures and sedimentary systems;
- Increasing the accessibility of geological outcrops to people of all physical abilities; thus outcrops located in inaccessible or dangerous locations become accessible;
- Public awareness amelioration and better understanding of science by creation of geo-referenced 3D interactive displays.

The use of GPS receivers in the field survey

1. How GPS works

GPS stands for Global Positioning System, which is short for NAVSTAR GPS (NAVigation Satellite Timing and Ranging GPS). The user of this satellite-based system can locate position fast and with high accuracy. Its initial purposes were military, and that was the reason for its development by the US Department of Defence which was initially controlling it. Later its use extended to scientific or even civilian purposes.

At first GPS may seem as a complicated system with equally complicated use, but the principle is quite simple. It consists of a constellation of 24 satellites (4 satellites in 6 orbital levels) orbiting at an approximate altitude of 20200 km every 12 hours.

Two carrier waves in L-Band (used for radio) are broadcasted by each satellite; these carrier waves travel towards earth at the speed of light. The L1 channel produces a Carrier Phase signal at 575.42 MHz as well as a C/A and P Code. The L2 channel produces a Carrier Phase signal of 1227.6 MHz, but only P Code. These codes are binary data modulated on the carrier signal. The C/A that is the Coarse/Acquisition Code (widely known as the civilian code), is modulated and repeated every millisecond; the P-Code, or Precise Code, is modulated and repeated every seven days.

A radio receiver is the device through which the GPS system works. This receiver acquires signal from satellites in order to locate its geographical position. Then the distance from the satellite is simply calculated by the GPS receiver, by measuring the travel time of the signals transmitted from the satellite and then multiplying it by the velocity (speed of light).

Distance = Velocity x Time

The GPS receiver computes its position and time by making simultaneous measurements of the distance of each satellite. At least three satellites are needed in order to define with precision a 2D position or a horizontal position. For the precise evaluation of a 3D position (latitude, longitude and height) at least four satellites are needed within signal range.

2. Accuracy

There has been a misconception about the accuracy of GPS. The US Department of Defence has intentionally degraded the accuracy of the system called Select Availability (SA) for many years; it was randomly degrading the accuracy of the signals being transmitted to civilian GPS receivers. However, SA was removed in May 2000. Therefore, there is now no interference to the accuracy of satellite signals, but accuracy is now based on the type of user device and its ability to eliminate error sources. The accuracy is affected by the following factors:

- *Ionospheric delays:* The ionosphere is the upper layer of the atmosphere ranging in altitude from 50 to 500 km. The particles which comprise it are mainly ionised thus causing disturbances on the GPS signals. The sun greatly affects ionospheric density; therefore there is less ionospheric influence during night time. The effect of the ionosphere also has a cyclical period of 11 years. For the current cycle, it reached its maximum in 1998 and its minimum in 2004.
- *Satellite and receiver clock errors:*

In each satellite there is a very accurate clock continuously monitored by ground stations (US Department of Defence). Errors of up to one meter can occur despite the presence of this equipment. Each receiver also has a clock but it is of course less accurate than the satellite's clock.

- *Multipath error:* Sometimes nearby objects, for instance tall buildings or lakes can cause the signal's reflection. Thus more than one signal may be received and therefore cause erroneous measurements.

- *Satellite geometry:* This means the relative position of the satellites at a specific moment. As long as the satellites are located at wide angles relative to each other, the possible error margin is diminished. On the contrary, when satellites are grouped together or located in a line the geometry will be poor. The effect of the satellites' geometry on the position error is called Geometric Dilution of Precision (GDOP). The components shown below, of which comprise the GDOP, can be individually computed but are not independent of each other. Additionally, in the case of low elevation satellite signals (anywhere between the

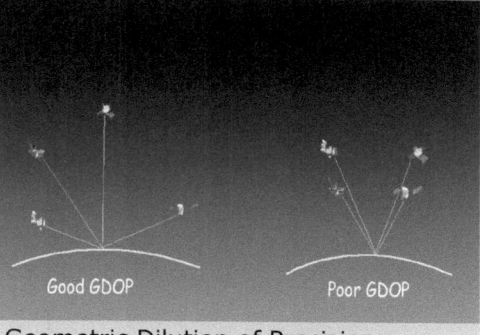

Good GDOP   Poor GDOP

Geometric Dilution of Precision.

horizon and up to 15 degrees above it) there will occur a longer ionospheric delay as the distance the signal has to travel is greater and thus the noise level is higher. In the more sophisticated GPS receivers an "elevation mask" can be set so that satellites below the mask are not used in computing position.

3. Types of GPS devices

Generally speaking, there are three types of GPS, with different levels of accuracy: Hand-held GPS or Navigational (accuracy = 1-10m), Differential Code-Phase GPS (DGPS) (accuracy < 1m), Carrier-Phase GPS (accuracy < 1cm).

- *Hand-held GPS:* The Navigational or hand-held GPS consists of a single receiver with the shape and dimensions of a mobile phone; it is affordable, comparable in price to a mobile phone, and very easy to use. It is the simplest GPS but also the least accurate. There is a frequent distortion by error sources which can degrate the accuracy of the position calculated from the satellite signals by several metres (about 15 to 100 m).

- *Differential Code-Phase GPS (DGPS):* This uses a differential measurement technique which

Hand-held GPS (12 channel, 0,3m post processing horizontal precision).

eliminates most of source errors, achieving results of sub-metre accuracy. This is a more complex system than hand-held GPS; therefore the device is more expensive. It consists of two parts: a base station and a rover receiver connected by a radio link. The base station, also called reference receiver evaluates the differences between the computed and the calculated range values by estimating what the ranges to the satellites should be after being located at a known point. These differences are known as corrections. These real time differential corrections are transmitted to the rover receiver (through radio) by the base station, and the rover receiver uses them to correct its measurements. The DGPS corrections are transmitted in a standard format specified by the Radio Technical Commission for Marine Services (RTCM). The Radio Beacon is a powerful radio transmitters. Set up around the coastline of many countries, these transmitters are located at old Radio Beacon stations, and have ranges of 100-150 Km. The frequencies used to transmit the DGPS signals are, in the old MF (medium frequency) Beacon band, around 300 kHz. These transmitters were initially used by marine navigators, but later in some countries, inland territories began to be covered by the system transmitters. Another radio transmitter is the OmniSTAR Inc, working in a way similar to that of the beacons. It consists of a network of GPS base receivers around the world, which broadcast corrections to user receivers. Access to these corrections is available by

Differential Code-Phase GPS (DGPS).

GPS uses a minimum of two receivers simultaneously. After an autonomous position is calculated using differential code methods, clock errors can be annulled by observing two satellites from two receivers by a method known as double differencing. Ambiguous results are resolved with the use of a statistical calculation of phase intersections from multiple satellites, once the better approximation of the position is known.

There are several measuring techniques that can be applied when surveying with Carrier-Phase GPS.

- *Static:* Used for high accuracy (about 5mm + 1ppm), measuring long distances. Data must be collected for several hours on two receivers simultaneously in order to achieve the best results. The duration of data collection depends on the length of the baseline between the receivers.

- *Rapid Static:* A form of static GPS which requires minutes instead of hours for satellite observation due to special ambiguity resolution techniques which use extra information. Accuracy can reach the centimetre on baselines less than 20km.

- *Real Time Kinematic:* This technique uses a radio to link so that the reference station broadcasts the data obtained from the satellites to the rover instantly. Baseline lengths are limited as data is transferred by radio, and accuracy will be in the range of 1-5cm. Nevertheless, it is evolving in the most popular technique since results are fast and co-ordinates are displayed in real time.

subscription. New satellite-based differential systems, free of charge, such as WAAS, EGNOS and MSAS, are also available. The Wide Area Augmentation System (WAAS) is used in aviation as it is designed to provide a higher confidence level in autonomous GPS positioning. The autonomous calculations can better define true position since WAAS corrects the atmospheric and orbital data, unlike radio and satellite differential. But since the system is designed for aircraft, there are still some limitations to non aviation users. Europe's first step into satellite navigation is the European Geostationary Navigation Overlay Service (EGNOS), which is an initiative of the European Space Agency (ESA).

- *Carrier-Phase GPS:* This differential system achieves accuracy ranging from centimetre to millimetre, depending on the measuring technique. The Carrier-Phase

Data is collected by most of GPS measurements techniques for post - processing, the exception being Real Time Kinematic Data collected by both receivers can be processed to obtain a better accuracy and/or to eliminate the noise caused by real-time operation.

4. GPS versus Total Station

Over the last decade, the Total Station Theodolite (TST) has rapidly become the preferred tool for surveying sites or undertaking topographical measurements, although frequently TST is the less attractive option when compared to GPS. Additional effort is required for the operation of a Total Station, and in many cases there are limitations:

- Where sites are remote or hard detail is poor, positioning may be unreliable.
- If a robotic system is not used, its use requires two people.
- Line of sight must be maintained between the instrument and prism.

On the contrary, there are many obvious advantages in the use of Global Positioning Systems:

- There is no dependency on permanent landscape features.
- There is need for only one operator for the survey.
- There is no dependency on a maintained line of sight between the base receiver and rover.

There are, however, some limitations with GPS that should be taken into account. The GPS receivers must always have a clear view of the sky in order to get signals from satellites. This is very important when the operator is in proximity to tall buildings, under dense forest, or when other interferences occur, because in that case satellite signal may be poor.

The use of handheld computers in field surveying

Implementing mobile mapping has significantly improved surveying efficiency. Many different types of devices may be used, such as handheld GPS receivers, palmtops and tablet PCs.

Laser Scanning for 3-D, 4D mapping

In the past 3 years, the introduction of terrestrial laser scanners in field surveying signalled a revolution. The technique has allowed rapid data collection of complex and complicated structures, both natural and manmade; before the introduction of terrestrial laser scanners this operation would have been immensely time consuming,

Hand-held computer (Palmtop).

and in some cases would provide less accurate models.

Surveyors and scientists find numerous advantages in laser scanners as a data capture technique. These include:

- Rapid non-contact measurement,

thus increasing productivity
- Increased data capture
- Integration of existing survey information with ease
- Health and Safety issues
- Highly accurate Digital Terrain Models (DTM's)
- Consistent and complete coverage over the desired survey area.

Not only are the data collected by geoscientists inherently in 3D, but the temporal dimension is also introduced. This obliges the geoscientists to develop the skill of four-dimensional visualisation of geological structures, in order to fully understand the datasets.

Despite this, the majority of field geoscientists still largely rely on paper-based mapping methodologies, whereby the 3D world is projected onto a 2D paper sheet. The paper-based environment is a 2D environment and therefore 3D or 4D relations that represent spatial and temporal relationships between different geological structures, are very difficult to represent and analyse adequately. So in order to use this traditional methodology, and in order to depict the 3D and 4D pictures that they have in mind, geoscientists must use corresponding diagrammatic model, e.g. a block diagram or "cartoon". This process relies on the geoscientis's skill and ability to form a realistic mental picture of the observed data and to be able to reproduce it in an appropriate form. This method has an obvious significant disadvantage: the models created during this process are not inherently 3-D, but simply involve a series of 2D sections or ortho-

projections that give the impression of a third dimension. Therefore any spatial information collected during fieldwork is effectively lost in the model building process. The original model remains inside the geoscientist's head and cannot be shared with other researchers because regardless of his skill, inevitably there will be a level of abstraction and simplification involved in the production of the final model.

A different strategy, for the exploration and investigation of potential hydrocarbon reserves, has been in use recently by geoscientists that work in the petroleum exploration and production industry. The rocks they wish to study are not usually exposed on the surface but are often buried beneath several hundred metres of ocean or rock strata. Therefore remote sensing techniques are employed to represent the sub-surface geological structure. In particular, high resolution (12.5m line spacing) 3D seismic survey data are collected that permit the construction of highly detailed and spatially accurate sub-surface models of hydrocarbon reservoirs at a resolution of 10's – 100's m. These models are not only spatially and geometrically accurate representations of the sub-surface geology, but they are fully 3D and can be viewed within an immersive environment by a number of people simultaneously.

This gives the ability to other geoscientists to share the "master copy" that is no longer locked within the mind of a single individual. Despite this and despite the ongoing advances in seismic surveying and data processing methods as well as in visualisation technology, data input from onshore outcrop analogues is still often required in order to provide information at a resolution below the current seismic threshold (20m). Heterogeneities can appear due to many geological structures and features (e.g. faults/fractures, vertical and horizontal faces variation) that lie at sub-seismic resolutions; these heterogeneities can significantly influence the characteristics of a hydrocarbon reservoir. The petroleum geoscientists, in order to introduce additional inputs into reservoir modelling parameters such as fluid flow, must rely on information (e.g. fracture spacing and orientation and faces variation) gathered from exposed onshore outcrop analogues. Output data and models that derive from traditional field mapping can provide information at a finer resolution than those that derive from 3D seismic data; nevertheless they represent mostly 2D samples with poor constraints within the third dimension.

Collecting information in office

Topographic maps

 All topographic maps are designed to be multipurpose. Representational design and map elements were selected in order to satisfy the users' requirements. For scientific applications, classic contour-based topographic maps, serve as a base for mapping and fieldwork. Topography is regarded as a continuous surface, landforms and features are mapped via variations of the topographic parameters. However, this surface can be represented by use of numerous techniques including the following approaches:

- *Perspective-pictorial maps:* This representation includes block diagrams that provide a view of a "block" of the Earth's crust from an oblique perspective, in which the top and two sides are presented. An oblique regional view is another perspective-pictorial map. Schematic maps are viewed orthogonally, with pictorial treatments of topography, stratigraphy, faults, and landforms. The physiographic diagram that relates landform graphics to geology and geomorphology is an example of schematic map.

- *Contouring:* Contouring is the mapping of a continuous surface using contours, or lines of equal value. The appearance of a 3D illuminated surface can be given by the contour lines symbolisation.

- *Hypsometric tinting:* This frequently used on wall maps approach is also called layer tinting, hypsometric colouring, and/or altitude colouring. The illusion of altitude change is achieved by the shading of areas between contour lines with colours that approximate the colour of land cover features. Generally, there is a gradual variation between colours on the map, which gives the impression that the surface change is continuous.

- *Hachures:* This approach uses lines that are positioned in the direction of greatest slope, such that the hachure's orientation is at right angles to contours. The use of lines of proportionate width in relation to the slopes' steepness (i.e., the steeper the slopes the thicker the lines), or of variations

Hillshade relief of Milos Island-Greece.

in line spacing in order to depict slope are also variations of the same method. Used effectively, hachures can give the illusion of an illuminated 3D surface.

- *Hillshading:* This approach depicts the Earth's surface as if illuminated by a remote light source. High relief regions are often displayed with use of hillshading, because it is effective in providing a very realistic depiction of topographic variation. The overlaying of other GIS layers (e.g., roads or streams) or images (e.g., digital orthophotos) in order to further increase the information content ameliorates the result. Hillshading can be combined with contours and/or layer tinting.

- *Morphometric maps:* Geomorphometric parameters of the topography contain morphological and some process-based information about the landscape and its landforms. Topographic parameters include relief, slope angle, slope aspect, curvature parameters, and degree of dissection. These maps can be viewed individually in order to associate topographic characteristics with surface processes and landforms, or "integrated" maps such as a "slope aspect" map can be generated to depict both parameters simultaneously. In order to better represent typical meso- and macro-scale topographic variations, additional scale-dependent parameters may also be computed and displayed.

- *Terrain unit maps:* In these maps the definition of landform regions takes place on the basis of descriptive terms, such as mountains, valley or hills. Structural topographic variations, such as highly dissected hill slopes can be the basis for other descriptors.

The predominant data set used in topographic representation and visualization is a DEM, which is one class of digital terrain models (DTMs). Others include triangulated irregular networks (TINs) which

DTM of Santorini Island-Greece.

represent facets on the landscape as non-overlapping triangular polygons. Regularly spaced grid cells with altitude values are the basic units of a raster-based DEM. DTMs should include geospatial referencing information with metadata for the map projection, altitude units, the map units, the datum, and the spheroid.

Transformation of the 3D surface of the Earth

Map projection is the means by witch the 3D Earth surface is transformed to a 2D map surface. The spheroid refers to the geodetic model used to capture the oblateness of the sphere due to polar flattening. Although they can be considerable on small-scale maps, spheroid-induced errors are small over the extent of most large scale maps. Datum is a set of numerical values serving as reference for mapping and defining a coordinate system. Elevation can be expressed in feet or meters, while map units refer to the planimetric coordinate system and are generally expressed in degrees or meters. The use of DEMs is generally straightforward, although there are some common errors and issues to be aware of. Among these are: missing data, poor edge matching, DEM production method sampling errors; also canopy, snow and ice elevations, rather than the ground surface elevation, can be represented by altitude values. Ancillary data sets and/or spatial interpolation are absolutely necessary in order to rectify missing data errors to any extent. Interpolation of edge pixel values as a mean of neighbourhood values is one of the existing solutions for recovering from edge-matching errors. Other DEM production errors include automatic scanning of contour maps at a resolution that produces striped DEMs.

The only solution in the case of altitude values representing canopy, snow, or ice, is the use of LIDAR or radio-echo sounding to determine the height of canopy and the depth of ice, respectively. LIDAR instrumentation is currently used to collect the highest resolution DEM data. Pulses are sent towards the Earth by aircraft-based LIDAR instruments and the transit time from pulse emission to pulse return is measured. Given that the speed of light is constant, transit time is a function of the aircraft's altitude above the terrain, measured along the LIDAR path. With dense scanning rates and the appropriate wavelength LIDARs can produce data containing returns from the first surface (i.e., vegetation canopy or building roofs), intermediate surfaces (i.e. ground vegetation), and, finally, the Earth's surface.

Software DEM analysis

There are various software packages that can process large DEM data sets. Many of these allow enhanced functions such as:

- *Landscape rendering:* Rendering software is used to generate simulated landscapes using concepts of selfsimilarity, periodic variation, and complexity. Consequently, these techniques can be based on fractal geometry.

- *Data draping:* A 3D view of a specific region can be created through the procedure of GIS layers, satellite imagery, and attribute information draping over a DEM. Three key steps are: depicting the surface, draping the theme, and setting the view parameters, such as model and camera positions.

- *3D fly-through animations:* Simulated flights through the landscape are accomplished by view changes, as the model and camera positions are modified. The value of this approach becomes obvious in mountainous environments where complex terrain, natural and geopolitical hazards restrict field access. Furthermore, this approach is necessary in order to obtain a comprehensive view of the landscape and is also essential for identifying high-altitude features that are not always visible from many vantage points in the field.

- *Subsurface modelling:* Enables rendering 3D representations of subsurface conditions. Spatial and temporal continuity, based on modelling programs that simulate water and material flow on the surface and subsurface, is achieved through use of 3D interpolation. Surface and subsurface conditions can be viewed from multiple directions (i.e., surface profiles, cross sections).

- *Geostatistical interpolation techniques:* Based on statistical models that address issues of spatial variation, such as autocorrelation and scale dependencies, in order to generate an elevation field based on sampled data.

- Fourier analysis and other geostatistical techniques.

Digitising topographic features

The use of reliable topographic information and maps is crucial for geomorphological field mapping.

Printed versions of topographic maps in their original sheet form, or in copies, are the basis of traditional field surveying; their number of colours is reduced in order to allow manual additions of point and thematic information.

A similar concept is followed in the map production process. Only one colour (usually grey) is initially used in the topographic base map which sometimes has a reduced content. Fully vectorised geo-data are quite versatile in digital cartography. The desired layers and objects can be easily selected by cartographers. Symbolisation (colour, line weights, line styles, sizes, symbols, fonts, etc.) can readily be changed. One source can produce many different maps because of the availability of fully vectorised data. Using data from different sources or in different scales may cause problems. In such cases map data must be manipulated and harmonised. Currently, topographic map data are being vectorised or have already been vectorised by most mapping agencies.

Data can usually be obtained in a GIS data format that encodes topology, while some cartographically symbolized data sets can also be used for desktop publishing applications.

The main topographic features included in a geomorphological map are:

- *Situation:* The situation elements (e.g., roads, paths, buildings) represent the portions of landscape altered by man and so represent an important part of each complete topographic map.

- *Areal features:* Only small symbolised area features are contained in topographic

27

maps, with the exception of forest areas (vegetation), area patterns representing swamps (hydrography) and in a few maps, settled areas (situation). Modern digital cartographic systems allow the easy inclusion of such features. Perfect natural resemblance cannot be obtained in a map (indeed, this must not be the goal of any symbolised map). However, the combination of relief shading, symbols, hypsometric area tints and aerial perspectives can lead to such results. Climatic zones, vegetation, and land cover, rather than hypsometry are the basic selection parameters for the use of tints which should resemble the natural colour of the terrain.

- *Hydrological features:* Hydrology on a map is represented either as point elements (springs, geysers, wells, etc.), line elements (rivers, creeks, canals, pipes, etc.) or area elements (sea, lakes, swamps, glaciers, etc.).

- *Contour lines and spot heights:* Contour lines and the spot heights are the most common topographic map elements. Carefully compiled contour lines with additional spot heights remain the most clear and vivid representation of terrain. The grouping of contour lines provides easy recognition and interpretation of morphologic features.

- *Relief depiction:* Analytical relief shading is the computer-based process of generating a shaded relief model using terrain and solar illumination geometries. The progress of computer graphics has eased the development of different methods for analytical shading that satisfy the particular

needs of cartographers. Generally, the magnitude of shading (grey values) depends on local topographic parameters, such as slope and slope aspect. The calculation of the cosine of the incidence angle between the surface normal and the light vector is used by an illumination model to define the grey value of each pixel. Aspect-based shading, calculated according to a modified cosine-shading equation, more precisely simulates the appearance of manual hill shading. A bright grey tone is used to represent horizontal planar topography in flat areas, since the slope aspect is undefined there. A combination of the areas with this grey tone and those modulated by aspect-based shading can be produced, using a mathematical function or an interactive control panel. The transformation of three components into weight for each pixel generates aerial perspectives. The first weight is the relative elevation of the considered point. The second weight is based on the exposure towards light, and the third is based on the relative position of a point on the hill slope. These weights are used to alter the calculated grey value. The exaggeration of vertical gradients leads to the suppression of relatively horizontal planar surfaces. For mountainous regions aspect-based shading and aerial perspectives are most suitable, whereas for lowland and flat areas the diffuse surface reflection approach is more suitable. A mathematical function can be used to combine of the two techniques in heterogeneous landscapes.

- *Digital cliff drawing:* Large parts

of mountainous areas are covered with solid rock or scree. Precise maps are required in those areas due to the presence of hazard mitigation. Techniques for clear and precise cliff representation were developed for the Alpine countries in the 19th century. In most cases, the methods used originated from slope-shading hachures.

Collecting information using remote sensing

The term "remote sensing" generally describes various techniques that enable scientists to examine sections of the landscape remotely. In all remote sensing techniques, output is in digital format, so various analysis and enhancement tools are provided. Aerial photography and satellite imagery, both photographic and telemetric, recorded at various wavelengths or bands of the electromagnetic spectrum are included in these tools for geomorphologists. The wavelengths used include several in the visible light range, several in the infrared range, and radar in the microwave range. Remote sensing categories can also include sound-wave studies, or sonar, and studies of regional variations on Earth's magnetic field. Great progress can be achieved in the geomorphological study of the ocean floor through use of side-scan sonar.

There are fields of remote sensing, such as magnetic, gravity and seismic studies, which are customarily not used in geomorphological research because the information they provide refers primarily to subsurface states; however the recognition and interpretation of surface phenomena can sometimes be assisted by knowledge of the Earth's interior (i.e., using gravity or magnetic anomalies to disclose the presence of a dike beneath a ridge, to relate to morphology of alluvial fans on which arroyos have been developed).

Aerial photographs and stereo - observation

Working with aerial photographs, geomorphologists have been able to view sizable portions of the land surface. Although individual aerial photographs arre spatially limited, a direct regional perspective of larger areas has been provided through the process of assembling groups of adjacent photos into a mosaic. The ability to perceive features not perceptible on site or on larger, more localised scales were often an important requirement in order to see landscape on a regional scale. Throughout the 1950s and 1960s, as new film emulsions became available after the refinement of remote sensing techniques, and as the new orbital imagery technology came on line, remote sensing became more and more dominant branch of geomorphological research.

Identification and mapping of geomorphic units using aerial stereopairs of the study area, is generally the first step in modern comprehensive geomorphological survey. Aerial photography, can help with the identification of the majority of the morphology, and provide answers to many questions on morphogenesis, such as denudation versus deposition processes. Slope angles can be estimated and classified, and the relative relief of all, but small features, can be determined with acceptable accuracy. Thus, the basic structure of the survey well outlined and mapped from aerial photographs, is provided to geomorphologists before field work.

Feld work is the survey's second step. The accuracy of aerial photointerpretation must be checked on site and measurements of small features must be made. Questions of morphochronology, such as the sequential history of terraces, must be ascertained. The basic pre-field framework of the study can be well provided by remote sensing techniques, using aerial photographs, but extensive field work is necessary in order to have a complete, accurate geomorphic survey leading to true detailed geomorphological mapping.

The final detailed geomorphological map is drawn after the completion of field work and the careful and thorough check of the preliminary map. It is generally multicoloured using complex symbology.

Satellite imagery

Orbital imagery, originally in the form of photographs taken by astronauts and then as telemetered imagery from unmanned satellites, has provided geomorphologists with a new perspective of the Earth's landscape. Initially, great hope was expressed for its value, but to date, there is no absolute proof of the supremacy of orbital imagery, over other techniques and methods, in providing new insights into the physical geography of our planet. During the last decade, orbital imagery of various types has been used by geomorphologists as the basis of geomorphological mapping with some success.

Gemini and Apollo astronauts have taken photos and Nimbus and ESSA satellites provided telemetered imagery, which both helped geomorphologists in their effort to develop geomorphological and geological reconnaissance maps of Lake Chad and of the Tibesti Massif. The conclusion for the use of imagery, whenever available, was that it helped in the discovery of very large features and structures which might otherwise have escaped attention. However, imagery needed to be of higher resolution if it were to provide new information not already available by other means.

In 1974, some of the first Earth Resources Technology Satellite (ERTS; now known as Landsat) imagery was looked, but it was concluded that its limited ground resolution would be only moderately useful for mapping purposes. However, a base map underlying other data, would be a helpful educational tool in reports, a source of information in defining "broad targets of particular interest and an tool providing inspiration for new ideas and promoting further thoughts about established concepts.

The limiting factor in the usefulness of satellite imagery was scale. The possibility of interpretation is limited to major relief features due to the small scale of orbital photographs. On the other hand, the combination of satellite imagery and supplementary field work could lead to the effortless creation of detailed and accurate geomorphological maps.

Nowadays, an area in which satellite imagery seems to provide a definite

30

advantage over aerial photography is in the territory of large-area small-scale geomorphological studies. New information and new mapping techniques can be provided as a result of the ability to perceive mega- and macro-structures, using mosaics of Landsat images. In 1981, space imagery was used for the construction of a world geomorphological map at 1:15,000,000 scale. The researchers were able to recognize all the information pertinent to a geomorphological map of such generalised scale, so they concluded that the imagery was quite satisfactory.

In 1982, it was found that there was no need to carry out a complete cartographic generalisation process starting from larger scale maps, since the direct development of a preliminary geomorphological map at a scale of 1:1,000,000 was permitted by the perception level of Landsat Multispectral Scanner (MSS) imagery. With adequate data from other sources, a standard geomorphological map of good quality could be developed. The direct regional mapping procedure was valuable since it saved time, effort and money; however it did not provide any exclusive information. Remote sensing technology has reversed the conventional pattern of geomorphological surveys, where studies were carried out locally and afterwards a regional picture was constructed. Now, one can select smaller areas for detailed work, after having done regional studies straight from aerospace imagery.

The use of satellite imagery in geomorphological mapping, particularly in small scales, has increased since improvements have been made in the resolution. However, it still is not clear how truly useful Landsat and other orbital imagery is in providing new data, not available from other sources. In areas where aircraft or ground investigations are not possible, because of the location's remoteness or special political conditions, the pictures are supplied by Landsat imagery. Pictures of areas difficult to photograph due to cloud conditions can also be provided by the use of radar imagery. Satellite imagery has much more to offer to scientists, to an extent only few have realised. At different scale levels, different types of forms are evident; that became clear through investigations of landscapes, carried out by geomorphologists, from larger regional perspectives. At the present time, most regional studies are viewed as generalisations from large-scale small areas to small-scale large areas. But regionalisation is more than generalisation. Regionalisation through generalisation is subjected to the risks of subjectivity. The process of generalisation must be decided at each and every level of generalisation by individual researchers, which means that a selection has to be made about what to include and what to eliminate from the analysis. An objective portrait with a regional perspective, relatively free from unintentional bias can be provided from space. Individual landforms and individual processes are placed into a wider geographic context with the assistance of regional view. As previously mentioned, the development of a single mapping key that satisfies all regions' requirements at all scales, has been impossible for geomorphologists.

A sample of QuickBird satellite image (Pan Sharpened – true color – 60cm). Santorini Island-Greece.

received from satellites like IKONOS and Quickbird, in terms of both increasing accuracy and enabling better recognition of large scale features.

The QuickBird satellite was set in orbit in October, 2001 and collects Panchromatic (PAN) images of 60cm analysis and Multispectral (MS) images of 2.4m analysis. The following table represents some of the satellite's principal features.

Orbit	Helio-synchronous orbit at 450km	
Analysis	60cm – PAN 2.4m – MS	
Coverage width per passage	16,5Km	
Scanning	Asynchronous (up to 750 lines/sec)	
Spectral channels	PAN: MS:	450 - 900nm Blue:450-520nm Green:520-600nm Red:630-690nm Near-IR:760-900nm
Image depth	11 bits per pixel	

The significant difference among landforms from different region and the requirement of a different cartographic symbolisation are a major cause for this. The proliferation of geomorphological mapping legends stems from the recognition of the complexity of geomorphological reality. The classification of individual geomorphic processes into series of basic categories such as tectonic, fluvial, or aeolian is relatively easy; however landforms are almost always the result of a complex mixture of processes interacting on the landscape. The multitude of often unique landforms that characterise the Earth's surface is the result of regional variations in the mix of geomorphic processes. Space imagery provides a direct perspective of the regional mix. A good view of the integrated dynamics forming the surface of our planet can be obtained through space imagery and the geomorphological mapping and analysis that derive from it.

Images with resolution up to 40cm can be provided by modern satellite cameras. The results of geomorphological mapping are enhanced by images of this kind,

A sample of IKONOS satellite image of the O.A.K.A. complex (Olympic Athletic Complex of Athens).

The IKONOS satellite was set in orbit in October, 1999 and collects Panchromatic (PAN) images of 1m analysis and Multispectral (MS) images of 4m analysis. In the following table are represented some of the satellite's principal features.

Orbit	Helio-synchronous orbit at 580km
Analysis	1m – PAN 4m – MS
Coverage width per passage	11Km
Scanning	Asynchronous
Spectral channels	PAN: 450 - 900nm MS: Blue:450-520nm Green:520-600nm Red:630-690nm Near-IR:760-900nm
Image depth	11 bits per pixel

The use of high resolution satellite images requires certain steps:

1. Image pre-processing

Normally, fusion between the panchromatic and multispectral imageries (at lower spatial resolution) must be accomplished. The Resolution Merge between them is performed by using the Principal Component method on specialised image analysis software. Moreover the standard type scene is only rectified and has to be orthorectified in order to be available for geological interpretation, by removing the effects of geometric distortion, caused by orography and manner of data aquisition.

Considering the high spatial resolution of 70 cm, accurate Ground Control Points (GCPs) must be collected and a proper Digital Elevation Model (DEM) has to be created.

2. Image processing

The DGPS GCPs and the DEM created from the topographic map are used for the orthorectification of the satellite data. The DEM was produced from: interval contour lines,digitised every 20 meters, spot heights and hydrography elements from the topographic map at scale 1:50,000.

The spatial resolution of such a DEM is approximately 30 meters and an RMSE of one-half contour interval is the maximum normally accepted. The accuracy of high resolution satellite images' orthorectification obtainable by a DEM is approximately ±10 meters, according to Kolbl (2001), which means that the created DEM could be used for the geometric correction of QuickBird images.

Then, initially, a rational non-rigorous method by means of the RPC file, the GCPs and the DEM is used for the image's orthorectification. The accuracy of the correction is approximately 5 meters, with a minimum of 1 and a maximum of 15 meters. A perfect orthorectification of the satellite image is not achieved by the conventional method, especially in non-flat areas; only the use of polynomials can moderate the lack of the basic full scene type and of the image acquisition geometry.

Thus, only the use of high order polynomial functions, can correctly orthorectify satellite imagery in respect to the terrain, a method called rubber sheeting. This

33

operation requires all available GCPs because high polynomial functions fit locally to the GCPs but not to the area between them. In this way the accurate assessment of position by checkpoints is rendered possible; while at the same time, internal imagery deformation is allowed.

3. Synthetic Stereo-Image creation

It would be better to use stereo instead of mono vision for the analysis of the geological and geomorphological characteritics of the area. Traditionally stereoscopic vision is based on a couple of aerial photographs or satellite images following the geometric characteristics of acquisition. One must create a second synthetic stereo-image on a PC, in order to obtain the stereoscopic view from a single QuickBird imagery.

Following the DEM, the synthetic image is created analytically by the introduction of parallax values directly proportional to the ground elevation. The following formula expresses the value of the artificial parallax (DP) for a single pixel of the image:

$$DP = Dh * K \ (1)$$

where Dh = elevation of the pixel above the minimum ground elevation, K = constant value determining the strength of the stereoscopic vision

4. Stereo-photogrammetry

After the creation of the synthetic imagery for the stereo vision, the choice is among the stereopair's hard print observation under a mirror stereoscope, the production of an anaglyphic image, and the orientation by means of a photogrammetric workstation.

Collecting information with digital image analysis

Initially a series of image enhancement algorithms is applied for the better identification of environmental and geomorphological data. Thus, the features to be studied are highlighted and distinguished from the other features contained in the image. This way, the influence of subjectivity on observations is reduced and the features that could escape one's observation, become apparent.

The basic methods of enhancement and correction can be summed in the following categories:

• Radiometric enhancement
• Geometric Correction
• Spatial enhancement
• Spectral enhancement

1. Radiometric enhancement

The radiometric type image enhancements refer to the configuration of Brightness, Contrast and Colour Density Slicing. By appropriate configurations it is possible to highlight specific image features, such as humid ground.

One of the handiest tools of depiction and configuration of pixel value frequency in an image is its histogram. Through the modification of the images' histograms, it is possible to maintain a homogeneous level. The data that defines the histogram may derive from the whole image, from a statistical sample, or from a specific region of the image.

Brightness-Gamma-Contrast: The simplest way of influencing the histogram, is by changing its brightness and contrast. The change of an image's brightness is performed by linear displacement of the pixels' spectral values towards values approaching white (255) or black (0). The change in brightness must not be great because many pixels get extreme values, which eventually leads to the loss of information. Particularly the Linear Clip algorithm causes a linear displacement of the values changing substantially the brightness of the image.

The adjustment of the gamma factor causes a non linear change of the image's clarity. This process provides a better clarity than brightness (it does not lead many pixels to the extreme values) and thus not much information is lost, as in the case of brightness. The equation which is used for gamma change is of the type output=input (γ). Of course, if the value of γ is equal to one, the image does not sustain any change. An increase of the value of γ causes significant change to the image's pixels with values close to 0 (black), while the pixels with values close to 255 remain almost unaffected. In order to restore the image to its initial form, the value 1/γ is attributed to γ (γ stands for the initially selected value).

As for the contrast, it refers to the degree of differentiation between the bright and dark pixels of an image. Change in contrast causes shrinking or expansion of the histogram to a lower or higher value range respectively. Low contrast is caused by the fact that different objects often reflect similar radiation amounts, in the visible or infrared section of the electromagnetic spectrum. Change in contrast can take place in a linear or non linear way.

The radiometric enhancement of the image under correction, takes place before its geometric correction, as the targeted control points are principally intense brightness alternations or stripes. The correct solution requires the insertion of many photoconstants from which the ones that present the greatest error are isolated and deleted, therefore the software gives a higher gravity to measurements considered more accurate. The errors and degrees of freedom are always presented in the photoconstants collection window. It is clear that the definition of the control point in the field should be made with the lowest possible error in position estimation (EPE-Estimated Positioning Error). The last stage after the geometric correction is accuracy control.

2. Geometric correction

Geometric correction is the procedure of introducing an image without geographic reference, to a system of coordinates or transferring it from one reference system to another. This process requires a reference system such as a map or another image, or points defined through use of GPS.

3. Spatial enhancement

Images or image sections characterised by small differentiations of brightness, are called low spatial frequency (low frequency) images. Respectively, if the brightness changes significantly within small distances, the images are called high spatial frequency (high frequency) images. Given the fact that spatial frequency describes the changes of brightness in an area, the methods of spatial information quantification that should be used are for example, processing of a group of neighbouring and non isolated pixels.

The spatial frequency of aerial photographs can be altered using two different techniques. The first involves the use of a spatial convolution filter that works with the application of convolution masks. The second is based on to the use of Fourier analysis that divides an image into sections of similar spatial frequencies by applying mathematical algorithms (Fourier transformation). This way, particular frequency bands can be modified, in combination or independently to others and finally compose an enhanced secondary image from the partial sections.

4. Convolution filters

The filter is a method of image processing applied in order to reduce noise, generalise the image features or locate the abrupt brightness changes. A filter can be applied to the entire image or to predefined sections of it. It is an array whose coefficients are selected in a way to highlight a specific property of the initial image. Arrays of this form are used for the calculation and comparison of the values of each pixel with its neighbouring ones, in a user-defined way.

5. Accentuation of edges

For the accentuation of edges the Laplacian filter can be applied. This filter applies an array of 3x3 dimensions, allowing the definition of a centre gravity coefficient for each neighbouring pixel.

6. Detection of edges

For the morphotectonic lineation detection the Sobel algorithm can be used to apply two 3x3 arrays in both neighbouring sides of the pixels. The value marked in the central pixel is the absolute maximum of the two results that occur by the application of the two Kernel arrays. Generally, the use of edge detection filters aims at the detection of abrupt tone changes, and also at their enhancement.

7. Enhancement by Fourier analysis

Fourier analysis is used to divide a monochromatic image into frequency groups. In these groups different algorithms can be applied depending on the anticipated result. For example it is possible to remove periodic noise, which is impossible to perform with the use of other algorithms. The divided frequency groups are composed to form the final secondary image.

Analysing samples in the laboratory

Sampling the open tubes for micro and macro fauna.

Instrumental measurements, using standard and formulated methods and analyses, are carried out on samples collected in the field. Necessary tasks may be: description of typical samples, classification on the basis of the Munsell chromatic scale, composition and description of the samples' stratigraphy deriving from geological / topographical sections or drillings, preparation for microscopical analysis, physicochemical analysis etc., depending on the nature and approach of the study. The selection of the methods and laboratory analyses that will be applied, as well as the selection of the protocol to implement for the purposes of the research must be framed by a correct series of designs, evaluations and speculations.

Introducing all information into a digital system

A given area, depending on its location on the Earth's surface, is best suited to a particular projection system. The projection systems simulate Earth by a three-

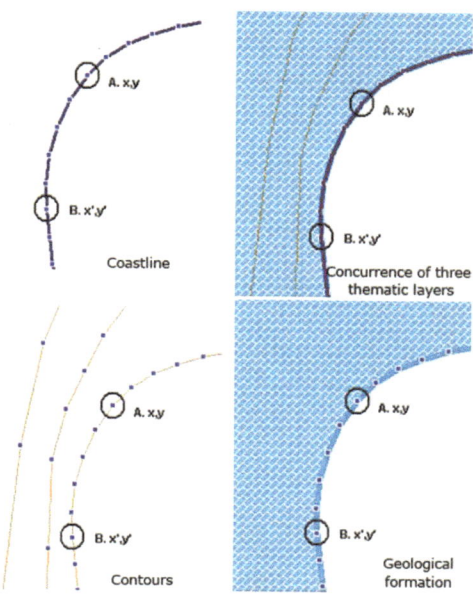

Coastline

Concurrence of three thematic layers

Contours

Geological formation

The coordinates of the points which define the coastline (i.e. A and B) coincide with the points of the zero altitude contour as well as with the limit of the geological formation.

dimensional spherical surface whose geometric features are known and which on the basis of these features, is divided in parallels and meridians. Through complex calculations the 3D spherical surface is transformed into a flat 2D surface. The sphere can be projected on a flat surface, cylinder or cone, giving respectively azimuthal, cylindrical or conical projections in a two-dimensional map.

There are many projection systems; some designed to cover the entire Earth (i.e. Longitude / Latitude), some divide it into zones (i.e. Universal Transverse Mercator, UTM) and others are limited within a country's area (i.e. EGSA '87 for Greece). Generally, a projection system designed for a specific area provides highest accuracy for this particular area than any other

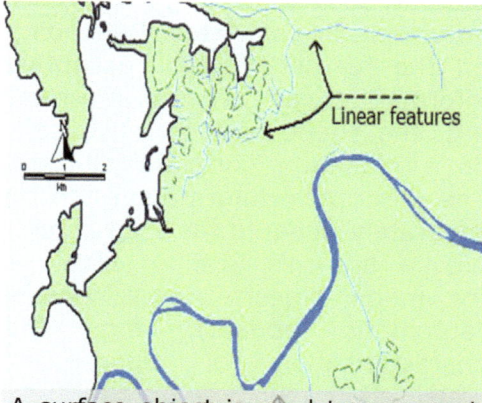

Linear features

A surface object is used to represent Acheloos River, while linear objects are used for the representation of torrents.

37

system covering the entire Earth. Depending on the extent of an area and the study's type, the appropriate projection system is selected.

GIS data can either be of raster, or vector format. Morphology is better expressed through raster data since they have the capacity to provide information for each grid cell, thus fully covering an area. A small island's drainage network that consists of torrents can be better represented by linear objects, while a river, such as Acheloos whose bed has a significant width, can be better represented by a polygonal object (surface). Swallow holes, due to their nature, can only be represented by point objects. Of course all of the above can vary depending on the data scale. (e.g. in a smaller scale, even Acheloos River is better represented by a linear object).

For every landform type or landform group an information layer is created, on which the objects will be digitised. Thus, for example, on a geomorphological map there will be one information layer for all the erosion residues and one that contains humid soil areas that will subsequently be distinguished in seasonal and annual. The latter distinction is performed through descriptive information, but the data are always found on the same information layer.

Data digitisation is either carried out from maps, satellite images, etc, or elements collected during field work. In order to use an already existing printed cartographic document, one must first georeference it. During georeference, points on the map are correlated to their actual coordinates in the selected projection system.

The scale used to view or process data, has no top or bottom limit, it is however delimited by the primary data. For example, once a map at 1:50,000 scale is used for the digitisation of primary data, one can project them even on a 1:1 scale; however these data cannot acquire detail or accuracy higher than that which they had at their original scale (1:50,000).

The snapping process is particularly important during digitisation since it ensures the concurrence of two or more objects. A topographical feature e.g. a coastline has often multiple properties; it is simultaneously the zero altitude contour, the limit of some geological formations or the edge of a port. This implies that this linear feature must be concurrent in all the information layers it appears. The crossing or concurrence of the linear objects is ensured through the snapping process.

When the digitisation is complete, the spatial objects that stand for landforms have already been inserted. Through their coordinates, the geographic parameters of each feature, such as area or length, are automatically calculated. Depending on the feature's type, descriptive information can be also inserted, e.g. the shape of a torrent's bed, or a planation surface's altitude. The descriptive information can be separately updated for each object, e.g. a branch's Stralher class or for many objects simultaneously (automatic update), e.g. the characterisation of dunes as stabilised in a study area.

The descriptive information is inserted in the data base of every information layer. Consequently, the

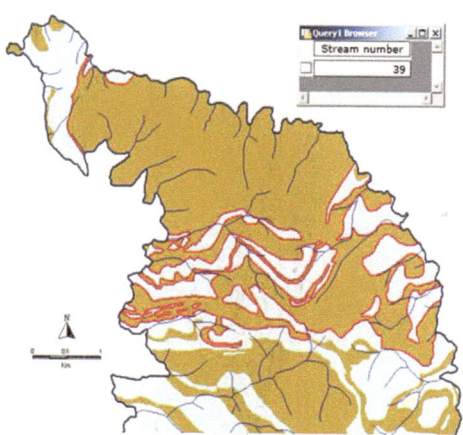

Thirty-nine torrents are intersecting the delimited geological formation.

network located within a particular geological formation can easily be located. Updates are made similarly, since it is quite simple to update each geological formation with the number of torrents that intersect it . If queries are based on spatial information then it is easy to isolate them; an example is the isolation of funnel-shaped dolines from oblate ones. Of course it is also possible to take a combined action, for example to select all planation surfaces of a specific altitude and update them by the number of drainage basins that intersect them.

design of the data base is rather important. Generally there are three field categories to be inserted in a data base, the string, the number and the special data. The special data are of True/False type, date type, etc. Very often a code is inserted in the database and is analysed in the information layer's metadata. The most common reason to do this is the large quantity of objects.

Once all the data are imported, a set of information layers concerning the landforms will be at the user's disposal. At this point data analysis, which is the actual use of a Geographic Information System, can begin. Analysis can be based on an object's spatial or descriptive information or their combination. An analysis can be simple, e.g. a query, or an update, or it can be composite, e.g. a model's application.

Queries based on spatial information can detect whether an object contains or is contained within another, the distance between two objects or help in the comparison between two or more objects. For example, this way, the shortest branch of a drainage

Through successive queries and updated information layers one can create layers with geomorphological features that derive from previous ones. Thus, based on contours, one can create an information layer of morphological slopes also containing aspect values, while drainage basins can be updated with values of drainage density and frequency. Models can even be created with the same techniques, for example an area's flood model. For such a model many information layers will be combined, as for example slopes, width of the torrent's bed, land uses of the wider area, atmospheric precipitates, etc.

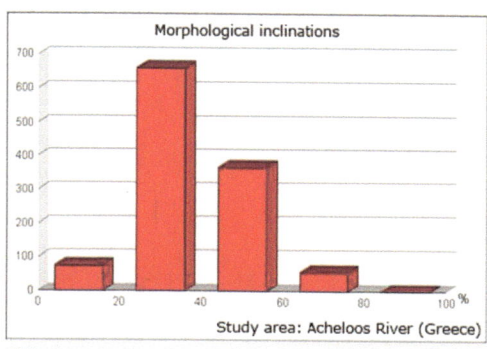

Histogram of morphological inclinations.

Among the most common types of analyses is statistical analysis. Most GISs have tools for calculating statistical parameters and for the creation of graphs. However, if the user desires, he can easily extract an information layer or the result of a search into another file form and import it in a purely statistical program. Of course, the possibility of importing the statistical results within the GIS is also available.

GIS has the capacity to spatially represent descriptive informations by using symbols for each geographic entity. Thus, a torrent's colour may change in reference to its class and its shading density may change in accordance with the value of the drainage frequency, in such a way that the spatial distribution of the descriptive information would be immediately understood. This feature is enabled within thematic cartography and four basic methods are followed:

1. Individual Values

The method of individual values is used to represent features with distinct values. The distinct values

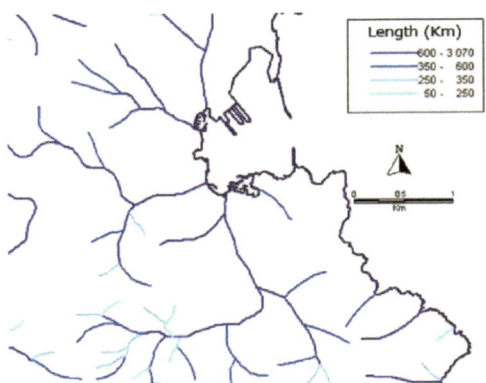

Grouping of branches on the basis of their length.

can be either string characters (i.e. geological ages), or numbers. Some examples of values can be the "1st Class" of a torrent, but also the value "Crystal limestones" of a geological formation. Depending on the descriptive information, the appropriate symbolism is selected.

2. Ranges

It is usual to group information in ranges, when expressed in continuous values, such as altitude, slope, direction, etc. There are many grouping methods, such as, equal ranges or equal width ranges.

Usually a drainage network's branches may be grouped based on their length and then each range is represented by a different tint. In the following example the branches have been grouped in a way that each range contains an almost equal number of branches. Through this method drainage density and frequency can be depicted.

3. Statistical mappings

This method is used for the comparison of values of one or more pieces of information. For example pixel density on a surface,

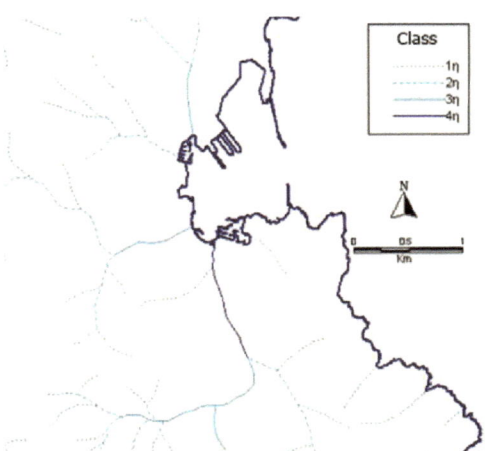

Representation of the branch class by different symbolisation and colour.

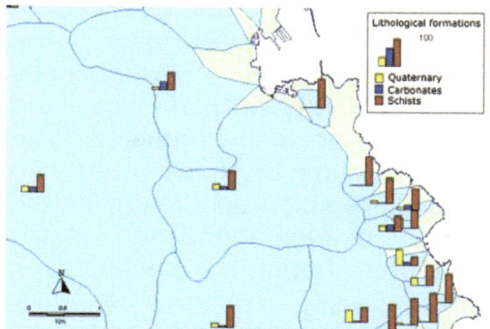

Branch grouping on the basis of their length.

image. Down-scaling DEMs results in exploiting more generalised landscape surfaces. For example, a smoother appearance and half the amount of detail will be provided by a DEM down-scaled from 1,024 x 1.024 pixels to 512 x 512 pixels. For the estimation of pixel values, when downscaling or otherwise transforming images, three interpolation methods exist, common to Photoshop and most GIS packages. The general, bicubic interpolation (default) works best with straight down-scaling. However, in case of a DEM's rotation, the use of the "nearest neighbour" interpolation would be more suitable and accurate, since this interpolation eschews anti-aliasing, and thus preserves crisp-edged pixels on the DEM's margin. Extremely smooth generalisation is alternatively achieved, by the application of the Gaussian blur filter to a DEM. The smoothing effects of Gaussian blur filtering are quite different from down-scaling; experimentation is required to achieve comparable levels of generalization between the two techniques.

increases proportionately to the multitude of settlements on a geological formation. Drainage basin graphs (histograms, pies, etc.), may represent the percentages of lithology that comprise it.

4. Simulation

In many cases, a piece of information is expressed by continuous values, but these are not known for the whole extent of the study area. Altitude for example is known along the contours lines and on survey markers. Based on these elements, a simulation of relief may be performed and values between contours lines can be estimated. The simulation result is a raster file with information in each grid cell. The most common file that occurs after a simulation of such kind is the Digital Elevation Model or DEM, where every cell has an altitude value. DEMs are also used for the extraction of information and files as the creation of topographic sections or visibility maps.

DEM enhancement techniques

1. Generalization

Down-scaling means decreasing the resolution or size of a digital

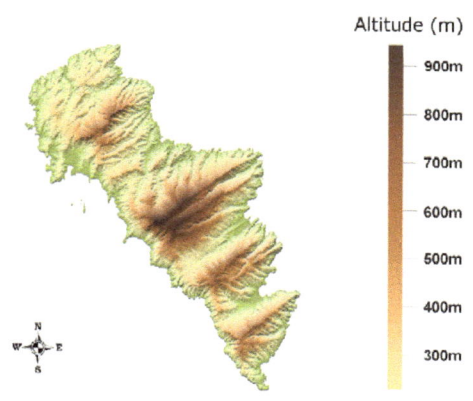

DEM of Andros Island-Greece.

Although generalization is most often applied globally throughout a DEM, it can also be applied in graduated amounts in order to achieve subtle visual effects. For example, the optical illusion of depth is created on a DEM, in a three-dimensional view, when generalisation is increased (and thereby visible detail is decreased) from foreground to background. Foreground to background generalisation also shortens rendering time, an important consideration when creating interactive environments.

Graduated generalisation can also be applied to the DEM's vertical axis, creating more detailed scenes at higher elevations than at lower ones. This technique follows the general point of view of the aerial perspective effect; a visualisation technique pioneered by Eduard Imhof that accounts for the veiling effects of atmospheric haze. Aerial perspective depicts highlands with greater detail and contrast than lowlands, because highlands are theoretically closer to the viewer, and lowlands are further away, enhancing thus three-dimensionality.

2. Resolution bumping

Resolution bumping is a generalization technique for manipulating GTOPO30 and other small-scale DEMs. This technique renders rugged, high mountains more legible and makes them look more natural when compared with unmodified data by the alteration of digital elevation surfaces.

When used for small scale 3D visualisations, unmodified GTOPO30 data typically produce mountains with a choppy appearance. Vertical exaggeration, a graphical necessity in small-scale, multi-landscape visualisations, exacerbates the choppiness. The mountainous areas of tightly packed ridges and valleys cause many problems. The representation of these areas can become illegible topographic detail, vertical exaggeration, and small-scale presentation are combined.

Down-scaling GTOPO30 data to a sparser resolution alleviates the problems outlined above. Patterns within mountain ranges are more accurately depicted by generalised data. Down-scaling elevation data, however, introduces new problems that are arguably worse than the (now-corrected) original problems. It alters the appearance of knife-edged mountain ridges rendering them excessively rounded, while simplified valley bottoms are displaced in 2D space.

The idea behind resolution bumping is simple: hybrid data are produced by merging low-resolution and high resolution GTOPO30 data of the same area; these combine the best characteristics and minimize the problems found in the originals. Two copies of a GTOPO30 file are used, one of high resolution and another one down-scaled to a lower resolution. These files can then be blended inside Photoshop by a proportional amount controlled by the user. This technique yields a new greyscale DEM that, if merged in the right proportions, combines the readability of the down-scaled data with all the detail one expects to find in mountainous terrain, without graphical noise. Resolution bumping in effect bumps or etches a suggestion of topographical detail onto generalised topographic surfaces. The resolution-bumped data create an elevated base in mountainous regions, from which individual mountains with diminished vertical scaling project upward.

3. Height manipulation

The raising or lowering of surfaces is achieved by lightening or darkening a DEM respectively, even with Photoshop's image adjustment tools (levels, curves, brightness/ contrast), when the DEM is later rendered in three dimensions. This technique can be used to modify vertical exaggeration globally over an entire DEM or, more interestingly, for selected topographic features. For example, a mountain hosting a ski area could be exaggerated in height above its surroundings.

Going one step further, applying lightening and darkening within selections can create simple topographic features. A volcanic cinder cone can be created by drawing a circular selection with a feathered edge and lightening the area within, forming a cone-shaped hill when the DEM is extruded in 3D. The cone has a more realistic appearance, avoiding excessive symmetry, if the circular selection is drawn with a slightly irregular shape. Finally, by contracting the initial circular selection by several pixels and applying a smaller amount of darkening, the summit is depressed.

Glaciers can be depicted by manipulating elevation on a duplicated DEM, positioned precisely below the original unaltered DEM in Bryce. In Photoshop, on the bottom DEM, an imported selection boundary representing the glacier's extent is drawn or imported and filled with lighter pixels. A domed effect, when the glacier is extruded in Bryce, is created by the use of a feathered selection boundary. By increasing the bottom DEM's vertical exaggeration and lowering its position in Bryce, the virtual glacier will protrude through the top DEM, neatly intersecting the valley walls.

Solid black can be applied on a DEM in order to form block diagrams and cut-away views, by taking elevation manipulations to the extreme. Black represents the lowest elevation value. When applied to a selected portion of a greyscale DEM, it flattens and lowers the topography to base level, the digital equivalent of a peneplain. The bottom-most elevation data can then be clipped from a DEM when rendered in three dimensions. This technique allows selected chunks of a DEM to be cut away, making cross-sectional views or revealing hidden features beneath the surface. Conversely, filling portions of a DEM with white abruptly elevates these areas above their surroundings.

On a large-scale DEM, filling small rectangular selections with white creates blocky shapes that, when extruded in three dimensions, can pass for primitive buildings (best done on flat surfaces to avoid sloped roofs). Text, point symbols, area patterns, and map linework can also be digitally embossed on topographic surfaces. This technique is potentially useful for developing tactile physical models, carved from DEMs by computer numerical controlled (CNC) routers, for the visually impaired.

4. Elevation flattening

The Gaussian blur filter is useful for more than generalising DEMs. A mathematical "soft" lens controlled by a radius slider that removes detail filters pixels (elevations), and is the "core" of the filter. Gaussian blur flattening, when applied to imported selection boundaries, yields benefits. For example, land water boundaries on DEMs often do not match the same boundaries on imagery or vector linework. This creates unpleasant misregistration near the shorelines when these data are later draped on DEMs. Editing the DEM solves the problem. By importing a selection of waterbodies taken from the geoimagery or rasterised vectors and applying maximum Gaussian blur, waterbody surfaces on the DEM become perfectly flat at their respective elevations in concert with the draped imagery. Obliteration of distinctive topography immediately bounding waterbodies occurs occasionally in this technique and is its only drawback.

Gaussian blur used in moderate amounts has other uses. Terraces uncannily similar to those created by actual earth-moving equipment are produced by elevation averaging when this is applied to a selected area on a slope. This is a useful technique on large-scale DEMs for depicting level areas around buildings. Moreover, excess height data from elevated protuberances are removed and data are added to bisected valleys, by moderate Gaussian blur applied to road selections, creating thus virtual road cuts and fills.

5. Painting

Edits may be made to DEMs by painting directly on their surfaces. Manual skills similar to those used in traditional illustration are required for DEM painting. In this way topography similar to natural appearance may be produced. Painting on DEMs is hampered by the disconnection between the appearance of the 2D greyscale DEM, the item that is painted, and the 3D model that will eventually be produced. The problem is especially acute when painting subtle tones that can be difficult to see on the monochromatic surface of the DEM.

Decisions about generalisation are required when painting on a DEM in order to depict temporal geological events through an image sequence or animation; these decisions are more difficult than those concerning single-image views. Nature is often much more complicated than convenient for illustration.

6. Topographic substitution

Comparisons with analogous present-day landscapes are often made in geology texts, when describing hypothetical former and future landscapes. This concept may be applied to the production of geological visualisations by cloning topography from one DEM to another using a technique known as topographic substitution. Topographic substitution is based on actual DEMs, so it is easier than DEM painting and looks convincingly realistic, providing that the user obtains appropriate DEMs. There are an unlimited number of options for mixing and matching topography to create hybrid landscapes.

The value of a geomorphological map in applied geology

The modern detailed geomorphological map provides a unique means of displaying all the

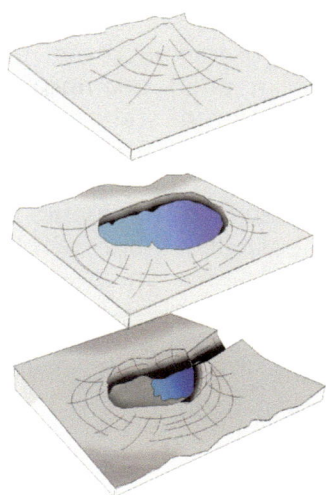

Comparison of analogous present-day landscapes.

various factors and features of the physical landscape in an orderly scientific fashion. This kind of map is the only analytical research tool developed so far, by which it is possible to approximate a portrayal of the Earth's complex surface and dynamics. It is scientifically valuable for research in theoretical geomorphology and likewise serves as a basis from which applied maps may be drawn, focused on special aspects of landscape, to support a variety of applied geomorphological researches.

A number of applications of detailed geomorphological mapping can be used by geoscientists in order to:

- get a precise picture of relief dynamics that enables the reconstruction of its development and helps evaluate the origin, factors and processes of transformation
- facilitate the search for spatial connections between landforms
- facilitate the development of comparative studies
- carry out a comparison between developed and developing landforms in areas of inconsistent or similar geological structure and under varying climatic conditions
- study the role of climate in shaping the Earth's surface by distinguishing types of relief according to climate

The complex nature of detailed geomorphological maps tends to limit their usefulness beyond the area of technical geomorphology and in most cases these maps are made by experts for experts. A genuine geomorphological map is an intricate document that can only be read by those with adequate specialised training. All of these factors tend to render their information inaccessible to those outside geomorphology. In spite of this, geomorphological surveys should constitute one of the basic elements in the preparation of most earth related projects.

In viewing geomorphological phenomena over a wide spatial context, as was required when preparing the IGU Geomorphological Map of Europe at 1:2,5 million scale, all researchers are subject to the discipline of working within an agreed international framework. The complex integration of the natural environment can be shown, for educational purposes, with the combined use of geomorphological maps and other physical maps. Finally, in relation to remote sensing, when mapping the landscape, an experienced geomorphologist can appreciate terrain types depicted on remote sensing images more easily and reliably than an analyst without such a background.

The special value of geomorphological mapping lies in its application to particular problems, by use of limited maps showing only the geomorphic features relevant to the particular question at hand. Such maps have either been derived by simplification of the detailed maps, or have been prepared using only the necessary data. Secondary maps are often more desirable because it is always possible to refer back to the detailed maps should further information be needed.

Geomorphological maps are of great value in the general field of Environmental Management, particularly during the planning stage. In 1974, geomorphological maps were found to be of principal utility at the initial field investigation stage of analysis by environmental organisations. They also considered the maps to be valuable as a basis for a number of special-purpose maps useful in various stages of environmental management. The resulting maps were simple and easy to read, showing only the information relevant to stability. The stability maps were developed from detailed maps drawn after a full geomorphological survey; this is a prerequisite to the development of maps useful for planning purposes. In a periglacial environment, microfeatures of the area such as patterned ground, solifluction lobes, meltwater channels, and eustatic strandlines could be mapped.

In structural engineering, geomorphological maps are used in planning, not only to deal with concerns related to construction in a difficult environment, but also to use the land in such a way as to enhance the community's aesthetic quality. There is great need for different types of applied geomorphological maps at different stages of planning and construction. Small-scale maps can provide regional analysis which would be valuable at the initial feasibility stage of planning. Large-scale or small-area maps would be valuable for questions of site investigation and could help in forecasting behaviour during and after construction.

Samos Island - Greece (by A. Vassilopoulos, N. Evelpidou)

Chapter 2

fluvial environments

fluvial processes

Rivers, water streams and fluvial processes

The drainage network is supplied with water in several ways: directly by the atmospheric precipitates, by the discharge of aquifers under the form of seasonal and permanent sources, by overflows and direct supply from the lakes, by side transfusion and infiltration of the underlying geological formations or by the melting of glaciers. Water is an important factor for the formation of the relief, and its role becomes more or less evident depending on the increase or decrease of its transfer and eroding capacity. Rivers have the capacity to erode, transport and deposit. Fluvial processes are related to the hydrosphere and belong to the exogenous processes that shape the relief, sometimes by acting constructively - land creation through deposition – and other times by acting destructively, resulting in lowering of relief.

The formation of a drainage network begins with the emersion (emergence) of an area from the sea. When the initial inclinations of the emerging land are low, erosion processes are limited. As the area continues to rise, erosion becomes more intense and fluvial/torrential deposits more abundant. If the elevation is continuous, the drainage network becomes deep and topography remains rough throughout orogenesis. The relief goes through a degradation phase when erosion rates are faster than elevation rates. When degradation phases last for a long period, the area becomes a peneplane.

Most of the descriptions of dynamic relief development processes in drainage networks refer to ideal conditions of uniform and linearly developing processes, in homogenous and isotropic rocks, under stable climatic conditions and with linearly developing tectonic movements. In reality, natural systems and geo-environments do not develop under these rules, but are exceptionally sensitive to the initially prevailing conditions (geological, tectonic, hydraulic, climatic, etc) and their micro- or macro- alterations that lead them to chaotic development forms, often non predictable.

One of drainage network's main feature is the drainage basin, which is the area drained by a branch of the drainage network. The line that defines water runoff direction between two neighbouring drainage basins is called a watershed. The tracing of a watershed begins from the lowest point of the branch, which is usually its junction with another branch and arrives again at the same point, so as to outline the area that is drained by this branch. During the tracing one must recognise that it goes through the topographically highest points (crest points), it intersects vertically the contour lines and it does not intersect the same contour twice. From all of the above it becomes clear that the watershed line never intersects valleys.

Torrential flow is that which appears seasonally in streams that have no permanent flow and have high water and sediment supply after periods of intense rainfall. The intense eroding activity of torrents is mainly due to the fact that they transfer sediments

Meanders and braided channels in the riverbed of Pinios River (Thessaly, central Greece), as it exits its mountainous path (by K. Pavlopoulos).

and suspended material of high density and volume and therefore increase their kinetic and erosional energy. The vegetation on the slopes of the drainage basin acts as an inhibitory factor to speed of water flow and contributes to the highest infiltration rates in the ground.

Every drainage network branch is characterised by the following: the drainage basin, which is the area where waters that finally reach the branch are accumulated, the riverbed, where material transportation occurs, and possibly an alluvial fan found in the main valley at the end of the branch's course.

As a drainage network gradually develops and approaches its base level, the branches' beds cease to be straight and form meanders. This means that they develop a sinusoidal, often repeated form that may be caused by one of the following factors: a) the presence of some obstacle located in the flowing and eroding course of the drainage network branch, b) the decrease of its flowing speed, c) the change of the transported material's composition (suspended, roundstones, boulders, sand, silt, clay, ions, colloids, organic material, etc), d) the resistance of the bed walls to erosion, e) the riverbank vegetation, f) its hydraulic load and hydraulic behaviour within its bed. If the bending of the meanders is intense enough, this leads to the formation of lobes which, as time passes, are cut off from the main bed and form U-shaped lakes (oxbow lakes).

The deposits on the meanders' banks are either natural levees, or sediments transversal to the natural levees. The latter are sand and silt formations along the meander's banks and owe their generation to flood periods, due to the decrease of transfer capacity. Frequently the natural levees are raised by man made structures in order to provide protection against floods.

Meanders are mainly developed in alluvial plains; however in cases where they are formed within valleys they are called incised.

Erosion speed and especially down-cutting speed, depends on various factors such as a) the speed of the waterstream, b) the volume of

Intense down-cutting erosion and gorge development because of the tectonic uplift of West Crete. Samaria Gorge (Crete, Greece) (by A. Vassilopoulos, N. Evelpidou).

water, c) the nature and abundance of erosion means (i.e. transported material), and d) the vulnerability of the rocks.

A drainage network consists of many branches, which are separated into classes. The primary branches of the drainage network are usually located in the highest altitudes of a drainage basin and are called first class branches. At the point where two first class branches join, a second class branch is created. The junction of two second class branches leads to the creation of a third class branch etc.

The shape of the drainage network reflects the tectonic, lithological and climatic conditions that prevail in the area. The principal types of drainage networks are the following:

- *Dendritic type:* The most usual drainage network type. The branches meet at random angles. It is usually developed in areas with geological formations of similar resistance to erosion and where there are no tectonic or geological structures to prevent or limit the branches' development.

- *Rectangular type:* It is developed in areas tectonically stressed by faults or joints that define the branches' direction.

- *Parallel type:* It usually appears in areas of steep relief or when the water flows through loose (non cohesive) materials.

- *Trellised type:* It is usually developed in areas where the geological and tectonic structure has an intense influence to the fluvial branches. The larger branches have orientations parallel to the tectonic structures (folds, rock contacts), while the smaller branches join the main ones orthogonally.

- *Disorganised type:* Has an irregular form and occurs in recently uncovered areas, after the retreat of inland glaciers.

- *Radial type:* The branches of the drainage network are developed radically and towards the outer section of an area; they are usually found in volcanic reliefs.

In addition to the above mentioned drainage network types, many more exist depending on the particular features of their drainage areas. Most areas present a composite drainage network that combines features of two or more of the mentioned types.

The drainage texture of an area depends on the frequency and density of the drainage network branches. Thus, a drainage basin is classified as one of fine drainage texture or well-drained when the density and frequency of its branches are high, and of coarse drainage texture or poorly-drained when there is low branch density and frequency.

The drainage frequency (F) is the total number of branches in a particular drainage basin divided by the overall surface.

F = ΣNu /Au

The drainage density (D) is the total length of branches in a particular drainage basin, divided by the overall surface of this basin.

D = ΣLu / Au

Valley slope inclinations often indicate the development stage of the area. A way to evaluate the inclination of a basin's slope is the measurement of the basin's contours total length multiplied by

Intense down-cutting erosion and gorge development because of the tectonic uplift of North Peloponnese. Vouraikos Gorge (Peloponnese, Greece) (by A. Vassilopoulos, N. Evelpidou).

the contour interval and divided by the basin's surface.

S = ΣLCu*CI / Au

ΣLCu = total length of the contour lines of specific interval.

CI = the specific contour interval

Network density is higher in impervious rocks and lower in more permeable ones. This happens because the rock's structure can directly affect the water runoff. In impervious soils, water runoff and stream development are favoured, while on permeable ones the water infiltrates and generates aquifers and water sources. An area's climate directly affects drainage texture, since precipitation type, temperature and wind increase or decrease the quantity of surface water of a drainage network at a particular moment. The climate also affects vegetation type and density in an area. Dense vegetation increases the soil's and rocks' infiltration capacity. The primary relief can also affect drainage network texture. Finally, time is also a determinant factor for the area's development.

Base level

The base level is hypsometrically the lowest area of a drainage basin, where the superficial water and part of the underground drainage water (except for the case of submarine water discharge that happens lower than the base level) discharges. Usually, the sea is the absolute base level, whose local average level comprises the altitudinal reference level. Within drainage networks and basins can be found local basic levels such as lakes, marshes etc. that have a major role in the development of drainage networks (sediment depositions, flow speed decrease, local network balancing etc).

All climatic changes in the geological past have had direct impact on the stability of the base level, followed as a natural consequence by the river's base level change. Thus, during the glacial periods of the Upper Pleistocene, when the sea level was low, the planet's geography presented a much different image. The continental shelves around the great continents were exposed and extended fluvial beds were created, for the drainage of the greater land areas towards the distant sea. Meltwater streams were forming before the glacier fronts and, overloaded with sediment, they expanded the already existent flood plains and created new ones. The valleys, due to glacial erosion, maintained the form of the capital letter U. During interglacial periods, when temperature was higher and the base level was higher than during glacial periods, the retreat of glaciers provided additional water supply rivers. Thus rivers gained more transfer capacity and erosive power. Of course, this description is only a simplified explanation of what happened. In reality many other processes also prevailed.

During the last glacial period and towards its maximum (20.000-18.000 BP), there was an important decrease in mean rainfall rates and the corresponding sea level lowering (-120 m to -130 m). During the period that followed, climate was more warm and humid, favouring intense erosion. Later on, the development of vegetation decreased water runoff for a period. Human activity followed, to define up to a point, the development of fluvial systems and relief. Of course all of the above should be considered as processes parallel to the endogenous ones.

Geomorphological development of valley systems

New water streams are considered to be those that have just begun their erosive activity and are mainly characterised by steep inclination and gorging erosion. Their erosive activity is stronger vertically than horizontally and their valleys are V-shaped. In loose material the V form is wide and broad, while in harder rocks it is narrow and deep. Erosion rates vary and depend mostly on the hardness of the rocks. Differential hardness of rocks, contributes to differential erosion; this results in the change of the streams' inclination and to the creation of waterfalls and torrents.

In case the river is old as far as its development is concerned, the relief of the drainage basin and its knick points have been smoothened. The called "old streams" old streams have low speed of flow and inclination. Erosion is more active horizontally, thus broadening the valleys, than it is vertically. In this phase the stream obtains a meandered form, continuing to erode the valley widthwise and forming a flat bed.

Gorges and canyons are typical forms of relief, whose formation is related to erosion. Gorges are very narrow valleys, created by streams in areas with intense discontinuities and faults, where rocks are easier to weather and erode. Canyons are deeper and longer compared to gorges and are created in areas where sedimentary rocks prevail.

The giant's kettles are formed at the foot of waterfalls where the formation of water whirls drifts of course-grained material. The friction of this material on the walls creates cauldron holes characterised by spiral wrinkles on the walls and roundstones on their base. Generally, the giant's kettles can either be vacant, or filled by the streams' depositions.

Transferred material deposits by waterstreams

When the solid supply of a waterstream is high and the transfer capacity is not adequate for the transportation of the corresponding solid material, flow speed is reduced and deposition of the heavier solid material begins. The decrease of a waterstream's transfer capacity may relate to the following causes:

• The increase of transported material without a proportional increase of its transfer capacity.

• The decrease of the stream's transfer capacity, because of the reduction of its water due to rainfall decrease or intense evaporation, or because of the decrease of morphological inclination.

• The decrease of the stream's speed and urge, because of the decrease of its inclination, the reduction of its waters or the broadening of its bed.

• Human intervention.

Speed reduction and therefore transfer capacity reduction of a water stream can occur either abruptly, or gradually. An abrupt reduction in transfer capacity, as in cases of abrupt inclination change (i.e. when a water stream coming from a mountain enters a plain) leads to the creation of alluvial fans. These deposits consist of sand, clay and roundstones, block the river's bed

A waterfall in Narada river (Canada) created because of a knick point (by C. Centeri).

and increase the possibility of flood incidents in periods of maximum water level. The water stream creates a wide, flat valley, called a floodplain, where it deposits fluvial sediments. When, in the period of maximum water level, the waters reach beyond its usual bed, the river begins depositing on its borders, thus creating natural levees. Principally in the arid and semi-arid areas characterised by intense changes in the fluvial systems (intermittent flow, torrential flows, etc) deposits have a typical radical form. These deposits, depending on the morphological inclination of the deposition area, are called alluvial fans or debris cones and are created in locations where transfer capacity is abruptly decreased.

Deltas are typical deposition forms created in river mouths, when the sediment provision rate by the river is faster than the removal and transportation rate marine processes. Deltas include a superficial and a submarine section. The delta section found above sea level is the deltaic plain, which is essentially the prolongation of the river's alluvial valley towards the

A V-shape valley in the Rocky Mountains (Canada) (by N. Tsoukalas).

sea and is also characterised by fluvial deposition processes. The section located below sea level is called prodelta and consists of very fine-grained sediments. The area of the two sections is different for each delta and depends on various factors, but principally on marine processes (waves, tides, sea currents).

Deltas can be classified according to the predominant process during the stage of their final formation. Each delta type corresponds to the predominant formation process: fluvial supply, wave activity or tideal activity. Most deltas do not belong strictly to one of the classes (Loboid, horseshoe-shaped, arcuate, etc), but their formation is a combination of several classes.

On the coasts of desert areas very large deltas, called exotic, are created. Their creation is due to rivers bearing a great quantity of waters, enough to traverse a desert area and deposit their sediment load in the coastal environment. The plains that are flooded by exotic rivers, as, for example, the Nile Delta, are exceptionally fertile and comprise the world's greater oases. These areas are of great financial importance and in the past they usually constituted important centres of cultural development.

Fluvial geomorphological cycle

The "cycle" refers to the successive development stages of weathering, erosion, transportation and deposition that may repeat during the life and activity cycle of a fluvial system.

Weathering stage

The study of weathering is an important part the study of an area's geomorphological development. Weathering has:

- An assisting factor (an agent) for erosion and material transportation. Weathering leads to rock's disintegration and fragmentation and contributes to the efficiency of erosion processes.
- A factor contributing to landform creation and development. Differential weathering, as a result of rocks' variety in an area, leads to the creation of various landforms, as, for example, tafonis, peeling domes, block fields, etc. Also, side debris is generated by weathering.
- A factor of the land's surface general lowering. In areas consisting of limestone, dolomite or gypsum, intense relief depression can be noticed, due to the soluble character of these rocks.

Weathering is divided into mechanical and chemical. The processes of mechanical weathering are generally divided in:

- Thermoclastic-Cryoclastic effects: high temperature fluctuations.
- Development of crystals: the salt crystals created after salt deposition by saturated water solutions, exercise tensions in the rock's pores and fissures.
- Dilatation due to discharge: this occurs due to rock's discharge when it has been eroded.
- Icecracking: this is due to the dilation of water within rock cracks and fissures when temperature falls.
- Organic activity: the organisms tend to open fissures, but also to produce substances that affect rock's composition and cohesion.

57

The biochemical processes of microorganisms play an important part.

- Colloid detachment: this is caused by soil colloids that exert tensions that can reduce the cohesion of the grains.

The most important processes related to chemical weathering are the following:

- Hydration: water absorption by mineral salts, resulting to the increase of their volume.
- Hydrolysis: occurs in siliceous rocks when water splits into H^+ and OH^-, resulting in the decomposition of silicates.
- Oxidation: oxides production when some minerals come in contact with oxygen in air or water.
- Carbonisation
- Solution: activity of the carbon dioxide dissolved in water.

Generally, the weathering process is directly connected to other processes forming relief during the fluvial geomorphic cycle, and for this reason, it is frequently difficult to distinguish the landforms that clearly owe their creation to weathering processes.

Erosion cycle

The erosion cycle consists of some typical stages of development of the relief:

Stage of youth

In this stage, the area is characterised by an intense relief of relatively big heights and pointed peaks while erosive processes are very intense. Every branch of the drainage network erodes intensely depthwise and creates a deep valley, with very steep slopes that converge to a V form. This stage is characterised by a general lack of

The waterfalls of Iguacu river in Brazil is the result of headward erosion by the river, up to old planation surfaces of less vulnerable rocks (by C. Centeri).

alluvial plains with the exception of those located along the main branches. During this stage the drainage network did not have time to develop and the interfluvial spaces are broad. It is possible that lakes and marshes exist in the interfluvial areas, in altitudes very close to that of a local base level. The watersheds are expanded and indiscernible.

Stage of maturity

In this stage erosion processes are continuing, but the peaks are more rounded and the slopes less steep. Valleys deepen further, more gorges make their appearance and alluviation is more intense. The area now, shows a well developed drainage network. The drainage network valleys are U-shaped, which means that the bottom is wider since erosion processes are more intense, not only depthwise, but also widthwise. The lakes and marshes that were probably formed during the youth stage have now disappeared. Furthermore, the alluvial plains cover an important area of the valleys' bottoms. The meanders have begun to appear in their alluvial valley.

Stage of old age

Erosion continues at a slow pace. The relief is lowered and is characterised by mild inclinations. If the erosion cycle is not disrupted, in this stage a valley with low relief will be formed and maybe also some hills, from more resistant rocks, residuals of erosion. These are products of differential erosion, while the plain that is formed is called peneplain. The peneplain is an area of very feeble inclinations; for this reason, the drainage network loses its transportation capacity as it enters the plains, resulting to the restriction of the erosive processes and to the amplification of the depositional processes. The valleys have broadened significantly and the alluvial plains cover large areas traversed by big meanders. The interfluvial areas have been lowered and the watersheds are no longer distinguishable. Lakes and marshes may possibly exist in the alluvial plains located at the base level or near it.

A peneplain is a surface similar to a plain, characterised by very small topographical fluctuation and created by the erosion that the larger area has sustained. Generally, it consists of materials from well developed fluvial basins. The creation of a peneplain is the last stage of the erosion cycle. The deposits of a peneplaine, are always considered to be more ancient than the deposits that cover them and posterior to the more recent layers that have been eroded.

A peneplain can be possibly be flooded by the sea, and then the continental erosive processes are partially substituted by marine activity. In this case the peneplain becomes a surface of marine abrasion.

Terraces

An area's erosional stages may be distinguished by so called "terraces". The terraces are large natural step-like formations that consist of a plane and a steep slope. Terraces, depending on their origin, are divided in fluvial and marine. Fluvial terraces are the remains of older plains that have afterwards been eroded by streams. This occurred due to

59

various causes as, for example, the change of a stream's transportation capacity or the tectonic movements in the wider area. In some valleys more than one terrace levels can be found indicating the area's different erosion stages. Older terraces are located higher than younger ones. The terraces are located higher than the river bed, do not flood and are very fertile; that is why they are often used for cultivation.

During interglacials, as ice was retreating upriver in the valleys and the vegetation coverage was regaining the area, sediment transportation was diminishing and streams unsaturated in sediment were eroding the area depthwise, creating terraces. Thus, at the time, the upper surface of every terrace had the age of the beginning of an interglacial. The older remains are located in higher levels, while younger material is located near the present level of rivers.

However, generally, in a chain of terraces, it is always the oldest that is located highest.

Disruption of the erosion cycle – Rejuvenation stage

The erosion cycle may be disrupted by a change in the base level, which can occur due to tectonic, eustatic or climatic changes.

Intense tectonic movements can lead to the change of an area's altitude. If the change is due to an upthrust, the area is lifted far from its base level, leading to the beginning of a new erosion cycle, since the river again erodes depthwise in its attempt to reach the new base level. Some indicators of the rejuvenation stage can be the knick points along the bed of a river that attempts to reach the new base level. Knick points are topographic step-like formations within the river's bed, created mainly because of differential erosion between the rocks of the area traversed by the river.

However, in cases of rejuvenation they may be the result of a lowering of the base level.

Climatic changes directly affect the erosion cycle and the relief type that will be formed. A climatic change in an area may disrupt its erosion cycle or begin a new one, resulting to the re-forming of relief.

A new cycle's beginning may be signified by the action of new waterstreams, the erosive activity of which is characterised by the vertical component more than by the horizontal. Again, down-cutting erosion results to the creation of V shaped valleys, narrower in hard rocks and wider in soft rocks.

The erosion speed of new streams varies in different areas and depends on the vulnerability of the rocks they traverse. In the youth and maturity stages, where relatively high inclinations are still predominant, streams erode at a faster rate and torrents or even waterfalls are created. On the contrary, in the old age stage morphological inclinations are smoothened, the hardest rocks have been eroded and waterfalls no longer exist. The fluvial system is characterised by the activity of the so called "old streams" that have low inclination and speed. The prevailing component of erosion is the horizontal one, which means that broadening prevails over valley deepening. Riverbeds cease to be straight and form meanders and micromeanders. The river may also present a multibranch bed. In this case there are two or more beds separated by alluvial islets. Usually, the multibranch bed is considered as an indication of very high solid supply, but this is not always true, according to the theory of Leopold – Wolman who suggests that this can also be an indication of a balanced river.

Davis stated the view that a river is considered to be in balance at the time, during its geomorphic cycle, when it just reaches the inclination that makes it capable of sediment transfer. When a factor changes in a balanced river, the system (as always happens in balanced systems) acts towards the direction that tends to absorb the change's effects.

A river can have balanced and unbalanced sections along its bed. The points between the balanced and unbalanced sections are not usually characterised by small changes of the river's lengthwise section, but mostly by abrupt inclination changes that form waterfalls and staggered beds. These riverbed sections with abrupt inclination changes are called interrupted profiles.

The basic question in interrupted profiles is to determine whether they occurred from the local resistance of the rock along the bed, or if they are the result of tectonic or eustatic movements that led to the displacement of the base level.

Intense down-cutting erosion and gorge development because of the change in the base level of the area. Matera (Italy) (by A. Vassilopoulos, N. Evelpidou).

main fluvial landforms

ALLUVIAL DEPOSITS

Material that has been transported and deposited by flowing water streams and rivers and mainly consists of roundstones, gravel and sand.

N. Peloponnesus-Greece (by K. Pavlopoulos)

ALLUVIAL FAN

Alluvial fans are deposits that look like radially extending debris cones, beginning where the drainage network branch leaves the mountainous area and starts crossing the plain. The creation of these formations is due to the decrease of the torrents' transport capacity. Alluvial fans vary in size, inclination and form of deposits, but they usually characterise tectonically active areas. There are alluvial fans the maximum deposition surface of which lies near the exit of the stream and fans the deposition surface of which has been moved downstream.

N. Peloponnesus-Greece (by K. Pavlopoulos)

ALLUVIAL PLAIN

An alluvial plain is the bottom of a valley which is fully covered by alluvial flood deposits, or by alluvial fans. Alluvial plains are produced by the deposition of the transported solid material of tributary branches, when their transport capacity and the slope have been decreased.

Samos-Greece (by A. Vassilopoulos, N. Evelpidou)

ARID VALLEY

In this case, the stream that formed this valley disappeared because of drainage or infiltration.

Erymanthos-Greece (by N. Tsoukalas)

CAPTURE

Capture of a branch of a drainage network by a branch of a neighbouring drainage basin. This is due to headward erosion, to changes of the base or relative flow level and to tectonic causes. The capture of a branch by another drainage network, due to erosion (rill or gully formation) is called *capture due to headward erosion*. The overflow and change of the flow direction of a branch towards a neighbouring branch and then elevation of its flow due to alluvial accumulation is called *capture due to inclination - overflow*.

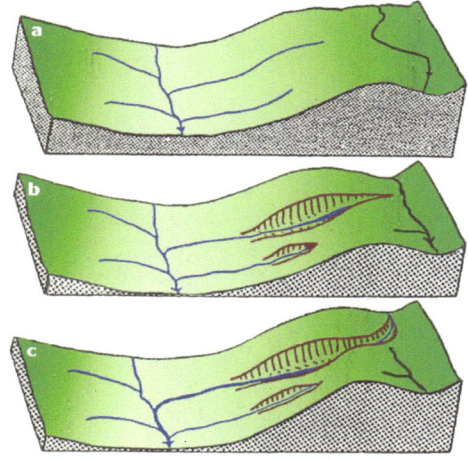

DEBRIS CONE

Debris cones are found either on slopes, or on the foot of slopes and precipices. Their formation is due to the decrease of a stream's transport capacity, during intense rainfall periods or afterwards, and consequently to the deposition of the transported solid material. On the top of the cones thick grained material is deposited, while towards the circumference the deposited material gradually becomes more fine-grained.

Vouraikos-Greece (by K. Pavlopoulos)

DELTAIC FAN/DELTA

Deltas are formed by fluvial/torrential material depositions in river mouths with relatively stagnant waters. Their presence represents the continuous capacity of a river to supply and deposit sand, silt and other weathering products at a much faster rate than the removal of material by wave activity.

2500 years ago, Herodotus emphasized that the alluvial area that is embraced between the branches of Nile and the sea has a triangular shape and therefore he used the letter Delta of the greek alphabet in order to define this area. Since then, the term delta has been modified in various ways, but is still generally used for areas where a river debouches without referring to the actual shape of the area.

Some of the principal factors that influence the character of a Delta are:

• The geological environment and the sedimentary sources of the drainage basin,

• The climatic conditions of the drainage and deposition basin,

• The tectonic stability of the drainage basin,

• The inclination of the river's bed,

• The river's regimen,

• The intensity of erosion and deposition procedures in the Delta area,

• The tidal range, the eustatism and the hydrological conditions.

The numerous combinations between the above mentioned factors and furthermore the time factor, lead to dynamic environment change within Delta areas, as might be anticipated since Deltas occur from the interaction of the procedures of creation (deposition) and destruction (erosion). The relation between the consequences of the mechanisms of creation and destruction, is based on the interaction of the intensity of the physical, biological and chemical procedures.

Contemporary deltas present quite a variety of size, shape, structure, composition, and way of formation.

Acheloos River-Greece

GORGE

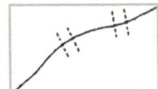

Very narrow and deep valley, with almost vertical slopes of a height greater than its width. Their formation is mainly due to the erosive activity of the water and the existence of faults or joints, that firstly facilitate weathering and then erosion. Genetically, its formation depends on sea level change, tectonic uplift and backward erosion.

Samaria-Greece (by A. Vassilopoulos, N. Evelpidou)

KETTLES, POTS (TORRENTIAL, OF GIANTS)

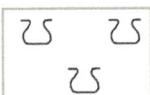

Shallow, sediment-filled bodies of water formed by draining floodwaters or retreating glaciers.

Glacier National Park-USA (by C. Centeri)

KNICK POINTS

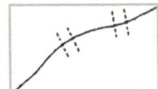

Abrupt topographic change in the bed of a stream or a part of the drainage network because of a tectonic line, or differential erosion. Waterfalls form due to knick points, in cases where the topographic change is large.

Vancouver Island-Canada (by N. Tsoukalas)

MAIN RIVERBED, FIELD OF FLOOD

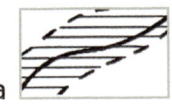

Channel bottom of a

river or stream whose margins, known as *river banks*, are confined by the normal water flow. During a flood stage, the stream overflows its banks and forms a *field of flood or flood plain*.

Olympic National Park-USA
(by C. Centeri)

MEANDER

Fluvial bed form characterised by the changing direction of the bed of a stream with asymmetric banks. In contrast to other sinuous bed forms meanders show symmetry. The concave section of the stream is steep, while its convex section is characterised by a small inclination. The meander form is due to the presence of some obstacle, located in the eroding course of the river. This obstacle may be a hard rock, more resistant to erosion than the ones surrounding it. If the meander bending is intense enough, then after a time period it may cut off from the main bed and form a lobe, which is a horseshoe-shaped lake. The meanders are developed mainly in the alluvial plains. If they are formed within a valley they are called embedded. The number of the meanders varies and depends on different factors such as, for example, the river size. The bigger the river, the more the number of meanders that its bed forms.

Free meanders (or wandering) are meanders of a small bed stream in an alluvial plain; these meanders change form and migrate quickly.

Meanders incised in the ground on which they flow are called *incised or incosed meanders*. They are not defined only by the water flow, but also by the combination of the valley with its bisymmetrical alternate banks.

Micromeanders are sinuous beds which are the result of the drainage micro-channels on a sloping or convex surface.

Aberdeenshire-UK (by A. Vassilopoulos, N. Evelpidou)

OXBOW LAKE (LOBE OF ABANDONED MEANDER)

A meander lobe which is cut off and abandoned by the main bed. Usually it is occupied by a lake or a marsh.

Acheloos River-Greece

PLAIN

Area of relatively small height and low relief,

surrounded by higher areas. Its material is of sedimentary origin and of recent age. The branches of the drainage network that cross the plains have a slow flow and meand form.

Hungary (by C. Centeri)

V SHAPE VALLEY

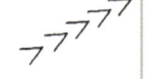

Narrow valley with great steepness whose form looks like the letter «**V**». The floor of the valley lies on the meeting point of its slopes. The down-cutting erosion defines its further development.

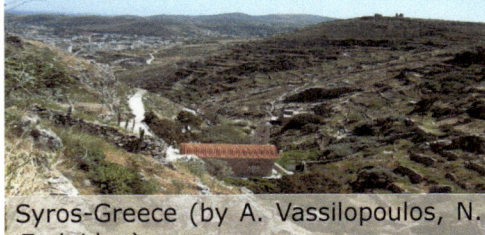
Syros-Greece (by A. Vassilopoulos, N. Evelpidou)

U SHAPE VALLEY

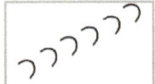

Valley whose form looks like the shape of the letter «**U**». The slopes of the valley range from concave to convex and are covered with colluvial sediments. This valley type is often met at periglacial areas.

Aberystwyth-UK (by C. Centeri)

⊔ SHAPE VALLEY (VALLEY WITH PLANE BASE)

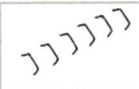

Valley with a flat floor which forms an alluvial plain between the two slopes. The width of the slopes ranges from a few meters up to tens of meters.

Loutraki-Greece (by A. Vassilopoulos, N. Evelpidou)

WATERFALL

Broken section of a stream's bed with continuous flow, characterised by an abrupt change of its topographic slope. The waterfall can be formed due to intense differential erosion, or to discontinuities (i.e. fault). The altitudinal change of the flow level in the case of a waterfall, can range from a few meters up to hundreds of meters.

Angel falls-Venezuela (by C. Centeri)

Vouraikos-Greece (by K. Pavlopoulos).

Sporades - Greece (by A. Vassilopoulos, N. Evelpidou)

Chapter 3

coastal environments

coastal processes

Sea water as factor of the coastline formation

Sea waters are important reformation factors for coastal relief. Mainly waves, but also tides have a significant weathering and erosive activity and create various coastal landforms. The material produced by weathering and erosion is carried by the waves to great distances depending on their transportation capacity level.

Wave erosion depends on many factors (e.g the sort of coastal lithologies and rocks). The main factors of the coastal formation are time, energy, sediment supply, change of the sea level and vegetation growth. Time guarantees the full dynamic counterbalancing after every change of one of the other factors.

The coast is constantly undermined and eroded by waves. As it retreats an abrasion platform is formed slightly tilted towards the sea. Materials produced by the weathering processes accumulate in deeper areas and form the -so called- continental terrace which, geomorphologicaly, is the natural continuation of the abrasion shelf. The abrasion shelf and the continental terrace form the continental shelf. The shelf's relief is characterized by gorges and channels which comprise the submarine natural continuation of the land's fluvial beds and continental valleys.

Retreat - Coastal Erosion

The waves erode the rocks of a

Coast with steep slopes. Rhodes (Greece) (by A. Vassilopoulos, N. Evelpidou).

coastal area not only directly, but also indirectly, creating thus cavities and notches on the rocks, thus reducing the coasts' resistance to sea erosion which will eventually result in their retreat.

The retreat of coasts which consist of hard rocks is an extremely slow process in relation to human time. On coasts with hard rocks, coastal notches or fissures are formed, which are broadened by the dissolvent energy of waves forming coastal caves. Coasts with an expanded coastal notches system have a form with multiple indentations which expand towards the land's interior.

On the contrary, coasts which consist of soft, loose sediments ,such as alluvial deposits, retreat at relatively fast rates, since the resistance to the sea erosion processes is more limited and waves with low transportation capacity can easily detach material from the coastline. In storm periods, a coastal retreat may occur, but the sand loss can be restored over long periods during which the waves have little transportation capacity. These cases are examples of temporary retreat of the coastline.

Waves

A wave is the expression of energy transmission from one point to another. The disrupted wave moves within the sea water (by diffusion) but does not sustain a permanent alteration as a whole. Many attempts have been made to classify the various types of surface waves, based upon their features. A more specific wave classification is:

- progression waves
- static waves

- free waves
- forced or violent waves
- deep water waves
- shallow water waves

In progression waves, every particle of the sea mass oscillates with the same amount of displacement and with the same period, but reaches its maximum at different time as the wave progresses through the mass. On the contrary, in static waves the displacement of each particle is different, but all particles reach their maximum displacement simultaneously.

The appearance and development of the sea surface waves depend mainly on the wind speed, the duration of the wind, the distance within a specific wave can be developed, and the initial sea surface conditions.

When the wind has a given constant speed, blows for a long time period and the distance is adequate for a wave to develop, balance is finally achieved, between the energy transported by wind seawards and the one consumed in wave breaking. This balance leads to the full development of the wave on the sea surface.

Wave features

The waves that are generated in water can be distinguished by the following features:

- Wave length (L): defined as the horizontal distance between two successive wave crests or troughs.
- Wave height (H): the vertical distance between a wave's higher (crest) and lower point (trough).
- Wave period (T): the time period

Beachrock formations at Kineta area (Greece) that go under destruction due to erosion processes (by K. Pavlopoulos).

required for two successive wave crests to pass by the same position and remains almost constant regardless of the change of other wave features.

A wave's speed depends primarily on sea depth in a proportional way. Speed refers to the basic wave component, however, in nature the wave consists of many components, which define the collective wave speed.

When a wave moves towards the coast, the water particles' circular velocity and particularly its horizontal component, reaches its maximum just under the crest. On the contrary, when a wave is directed towards the open sea, the circular velocity of the water particles reaches its maximum value just under the trough.

In a troubled sea, it is difficult to evaluate precisely the wave's height and in order to do so the substantial wave height is used; this height is the average of the one third (1/3) of the highest waves of the total wave range. For the coastal zone, and since the waves are breaking, the height of their breaking is used as wave height.

The wave energy depends only on the wave height and is independent of its other basic features.

Coastal currents

Coastal currents are those that are created when waves approach the coast. These currents, depending on the features of the waves that create them, may transport sediment to and from the coast. The coastal currents are the most important cause of sediment displacement along the coastline. The continuous arrival and breaking of waves on the beach leads to the accumulation of sea water mass. The discharge of this mass is effected by the creation of currents that move either parallel to the coastline, or in an off-shore direction. The type of current that will be formed depends on various factors, the most important of which are the angle of wave incidence on the coast, the morphological characteristics of the coastline and the morphology of the submarine relief.

If the waves' incidence on the coastline is vertical or almost vertical, then a kind of cell circulation is generated due to longshore currents and rip currents. If the waves' incidence is of a different angle, longshore currents are generated. The activity of these longshore currents is limited to the area in front of the wave breaking zone. The particular features of the longshore currents depend on the angle under which the waves approach the coast. Their speed ranges from a few tens cm/sec up to 1m/sec.

Rip current activity leads to

sediment transportation from the coast towards the open sea. Their particular features depend mainly on sea level rise, due to the accumulation of a water mass in the wave breaking zone. The rip currents are strong, narrow, their beginning lies at the wave breaking zone, and are directed towards the open sea. Their length can reach 60-750 m, their speed is higher than 50 cm/sec and they can often exceed 2 m/sec.

A gentle slope coast at Marathonas area (Greece) which consists of a variety of coastal materials, such as gravels and coarse sands (by A. Vassilopoulos, N. Evelpidou).

Sea currents generation is due to various factors, principally:

- The wind: An important factor since, apart from taking part in the generation of waves, it also carries away surface water masses towards the direction it blows.

- The tide: Another reason for current generation, this is of little importance for the open sea basins, but when taking place inside closed basins of characteristic morphology (Straits of Euripus, English Channel) it can possibly produce very strong currents, during low and high tide phases.

- Hydrostatic pressure variations: Sea currents are also created due to the presence of different density values that cause the displacement of the more dense mass towards the area of the less dense one.

- Earth's rotation: This factor affects sea currents' course and development and is expressed by the Coriolis force.

It is therefore possible that, during the movement of sea masses, more than one of the aforementioned factors participates, or that other parameters of secondary importance take effect.

There are four principal current systems resulting from wave activity on the coastal zone:

- A closed circulation system that consists of rip and longshore currents.

- A system of coastal currents originating from the angular incidence of waves on the coast.

- A system of deviational currents. If the wind blows for a certain period of time, towards a constant direction, it carries away molecules of the surface layer and the movement gradually expands towards the bottom. If the earth was static, the deviation current would have the same direction as the wind, but the Coriolis force, which is caused by the Earth's rotation, forces the superficially developing current to diverge by 45º to the right on the northern hemisphere and to the left on the southern.

- A System of inclination currents that is the consequence of deviational currents. In reality, when one of these currents produces water accumulation towards the coast,

Steep coast in Dunnottar Castle (Scotland) (by A. Vassilopoulos, N. Evelpidou).

the accumulated waters have the tendency to roll in the opposite direction, due to the generated inclination. The direction of the inclination current should be opposite to the one of the deviation current, but the Coriolis force creates in this case too a deviation of the current, whose direction is vertical to the coast and is also directed to the right on the northern hemisphere (left on the southern).

Sources of coastal sediments - Balance of the coastal zone sediments

Coastal landforms are formed by material produced from rock weathering and erosion. This material is transported to the coastal zone by water (rivers, torrents, glaciers) or wind.

The formation of coastal landforms (sea shores, dunes, berms, beach cusps, etc) is due to the processing and redistribution of coastal zone sediments by various energy forms acting on a coast. Energy in the coastal zone is expressed through the activity of waves, tides and sea currents.

Erosion that takes place on the coastal zone is responsible for a very small percentage of the sediments that enter the sea. In 1960, it was discovered that, even in temperate areas where wave energy is more powerful, less than 5% of coastal sediments are the result of erosion of coastal cliffs. This deduction was later supported by other researchers as well. In 1978, it was evaluated that an average erosion rate of 5 cm/year for the whole of world's coastal cliffs, (almost 50.000 km in length), would provide only 0,04% of the full amount of sediments supplied to oceans by rivers.

Rivers and torrents provide more

than 90% of the sediments that reach the oceans. The next most important sediment sources are the glaciers and finally the biota.

The sediment, which is transported in various ways, does not directly enter the coastal zone. On the contrary, it participates in a large scale sediment budget. Sediments move between two places of sediment accumulation, the continental shelf and the various coastal deposits such as sea shores, dunes, and river mouths.

Sediment displacement from very deep areas to the shore is mainly caused by tidal currents, or swell waves (waves during a storm), which can reach the necessary speed for sediment transportation over the sea bottom. In shallow waters, waves and coastal currents created by wave action have the predominant role. Offshore sediment movement can occur during storms and can also be performed via individual "paths" such as transportation along the coast leading sediment to areas of great depth. Furthermore, sediment transportation from the coast to great depths can be achieved through submarine canyons.

The interaction between sediment storage and sediment transportation can occur in a very short time-span, when during the summer swell waves move sand towards the coast, or in a larger time period such as the sequence of glacial and interglacial periods.

Knowledge and understanding of the coastal material's origin and of its transfer mechanisms is necessary for studies concerning an area's coastal geomorphology or the execution of coastal works.

The coastal zone sedimentary budget is the result of the action of several land and sea processes, which are divided in two main categories:

• The ones that bring sediment to the beach.

• The ones that remove sediment from the beach.

A coast's progression or retreat is determined by which category is predominant. In the case where the contradicting forces are equivalent, the position of the coastline remains stable. Anthropogenic structures, such as residential and touristic settlements along the coast, hydroelectric and irrigation dams as well as anti-erosive works for the protection of soil from erosion, have led to the reduction of land material supply.

Coastal sediments balance

Sediments are moving between the two principal areas of deposition otherwise characterized as "sediment depots" which are the sea bottom and the coastal zone.

In a study of sediment transportation along the coastline, it is important to determine the lateral borders of the coastal zone's section, where the quantitative evaluation of the sediment supply or removal factors may be needed, so that no factor is underestimated.

Special investigation must be made of possible human constructions on the coastline, as well as of the river and torrent mouths, even if they are located far from the study area. When these mouths are located near coastal cliffs that consist of non-cohesive rocks, their erosion provides a significant amount of sediment to the coastal system.

Coastal zone sediment balance

Sediment supply	Sediment removal
• Fluvial supply of solids (sediment transportation by rivers and torrents) • Coastal cliffs erosion • Sediment transportation by sea • Sediment transportation towards the coast by wind (Aeolian transportation) • Biogenic deposition • Artificial enrichment, rambles (human activity)	• Coastal transportation • Off-shore transportation • Sediment transportation away from the coast by wind (formation of coastal dunes) • Entrapment and removal of sediment through undersea canyons • Sediment removal due to human activity, (i.e. sand, gravel)

Sea level changes

The coastline is constantly changing through time. Its development depends on a series of non-linear factors such as vertical tectonic movements, hydro-isostatic movements, climatic conditions (atmospheric pressure), tides, waves, sedimentation, aeolian processes and human activity. It is obvious that the creation of a mathematical model, both for coastline and sea level change for the past and the future is particularly difficult because of the multi-factor variables and the chaotic conditions that are developed.

For the representation of the coastal paleo-environments and the sea level changes, a series of "absolute" dating methods (^{14}C, OTL, Pb, U/Th etc.) is combined with micromorphological (sedimentary) and micropaleontological sediment analyses. The dating methods are applied on sediments (e.g. peats), shells, archaeological findings from within sediments and on the adhesive material (cement) of coastal and submarine beachrocks. These results are used in geomorphological and morphotectonic analyses and on the same time validate the data on the paleogeographic development of a particular area. The most common landforms to be used as "indicators" of sea level changes are: a. beachrocks, b. notches on resistant rocks c. sea platforms d. biological indicators corresponding to marine organisms that lived close to sea level (a few centimeters above or below i.e. Vermetidae, Lithodomus and corals).

The forecast concerning coastline and sea level future changes is done with the combination of remote sensing data by satellites (Topex Poseidon, Jason, etc) with date-series of seasonal changes during the last decades, measurements of tidal ranges on global or local scale and mathematical models principally based on climatic changes (temperature rise on global scale).

The combination of scientific methodologies and approaches on a shared database for a particular area, may improve mathematical simulations and scientific prediction scenarios for sea level changes on local and global scale, so as to simulate reality in the best possible way. The results of this combinatory scientific approach may constitute valuable tools for the planning and realisation of decisions concerning an integrated, viable development of the coastal zones.

Sea level changes during the Upper Pleistocene

Sea level has changed several times in relation to the current level. During the upper Pleistocene four glacial and interglacial periods occurred resulting in global scale sea level changes.

1. The astronomical theory (about glaciers)

This was first introduced in 1864 by the Scotsman James Croll. Today it is widely known as the "Milankovic theory" named after the Yugoslav astronomer who improved this theory, in the 1930s. In the 1980s, it was proved that glacier eras are closely connected to changes on the Earth's axis orientation while moving around the sun.

The change of the Earth's axis direction is a complex combination of three separate movements. By combining these three movements, one can find in which areas, particular sections of the Earth receive less solar heat, where glacial processes are more likely to happen.

• 1st movement: *The precession of the Earth's rotation axis*, according to which the trace of the axis forms a circle within a period of 19.000-23.000 years. This happens because of the gravitational effect of the sun and moon, on earth's equatorial bulge. Its effect cannot

be traced in short-term variations such as the seasonal change within a century, but only over a long time-span, in periods of thousands of years. It is defined as the change in the orientation of the Earth's rotation axis in relation to its orbital plane, with a period of 21.000 years. This period is referred to as precession cycle of the rotation axis.

• 2nd movement: *Obliquity of the Earth's axis* (it is also defined as axial tilt). The angle between the Earth's axis and the line vertical to Earth's orbit plane (plane of the ecliptic), is slightly reducing and afterwards increasing within a time period of 41.000 years. The difference between the two maximum displacements is small, approximately 3 degrees (from 21.8° to 24.4°), but is enough to change the amount of the solar energy that reaches the Earth's surface. Today the angle is approximately in the middle (23,4°) and reducing. Therefore, we experience small temperature variations between winter and summer.

• 3rd movement: *Eccentricity:* Earth's orbit around the sun is not circular but elliptic. The orbit is parameterised by the eccentricity (e) which is based on the ratio of the divergence between the maximum and the minimum diameter, divided by the aggregation of the maximum and the minimum diameter of the ellipse; thus, a ratio of 0 indicates that the orbit is a perfect circle. The periodical changes of the earth's eccentricity have a frequency of 100,000 years. Thus, every 100,000 years, the Earth's orbit around the sun changes from mildly elliptic (e = 0.058) to almost circular (e = 0.005). Change in eccentricity occurs because of the gravitational effect other planets of the solar system have.

Pleistocene sea level changes

The Pleistocene began 1.8 million years ago and is known as the era of glaciers since rather low temperatures prevailed during this period, in comparison to previous geological eras. During the Pleistocene, 17 successions of cold (glacials) and warm (interglacials) climatic phases have been documented. The glacial periods lasted for approximately 100,000 years, while the interglacial phases lasted around 10,000 years. The last glacial began 70,000 BP (years before present) and ended 11,500 BP. During interglacials the ice blocks were melting and the sea level was rising whereas the opposite happened in glacials. Between glacials and interglacials, intermediate stages of warm and cold periods took place, with duration of almost 1,000 years. These small successions were causing eustatic changes. Along the coastlines, indications of the sea level during interglacials can be observed, in the form of elevated coastal platforms, of coastal notches or coral reefs. A detailed sedimentologic and stratigraphical examination is required in order to reveal the earlier phases.

Sea level changes during the Holocene

At the end of the last glacial period 18.000-20.000 BP the sea level was about 120m lower than it is today. By the end of this glacial period and the retreat of glaciers, the sea level was initially rising fast, 1m/y (meter/year) up until 6000 BP while afterwards the rising rate was reduced to almost 2mm/y. This is generally acknowledged and has been proved by a series of dating processes and through sedimentologic, morphotectonic, geomorphologic and archaeological analyses, on a global scale. These rates are not steady, especially in tectonically active areas, like Greece, where in certain coastal areas vertical tectonic movements do not conform to these figures (Falasarna, Western Crete +7m the last 2,500 years, Manika, Euboia, -4.5m the last 3,400 years), as it is confirmed by archaeological findings.

From time to time, various researchers have suggested curves depicting the sea level changes, trying to isolate the tectonic factors and trace a curve which will principally respond to eustatic changes.

In 1973, after using the data of three tide-measuring stations for the year 1680 AD, it was discovered that between 1780 and 1850 AD the sea level was depressed due to a new spreading of glaciers (Neoglaciation).

Classification of coasts

Generally

Each coastal type reflects the dynamic processes and the potential of its marine and land environment. The various types of coasts depend on their material, their typical geomorphology and the way they were generated.

Erosion processes taking place at volcanic rocks coast. Aigina (Greece) (by A. Vassilopoulos, N. Evelpidou).

The coastal material might be homogenous indicating a high energy environment, or irregular indicating a low energy environment, related to intense sedimentation, i.e. a coast with landslides. The presence of roundstones may also indicate possible feed of the coast by a torrent. Silt and clay material could indicate the interaction of the coastal system with a deep submarine basin, since fine material can be easily transported by waves and currents at long distances.

From time to time various classification attempts have been made intended for coasts. However, none of the classifications can be considered as entirely successful. This mainly happens on account of the fact that a category of classifications may focus on coastal generation (genetic classifications), while another on coastal description (i.e. cliff coasts, deltaic coasts etc).

Classification according to Shepard (1948)

A classification was proposed by Shepard in 1948. According to its more recent version, coasts are divided in:

I. Primary coasts are those whose formation is the result of non marine processes. Primary coasts are further classified in:

   A. Coasts of overland erosion, created by the erosion of the land's surface and afterwards flooded by sea, when its level is rising:
      1. Fluvial valleys flooded by the sea. These are river mouth systems of relatively small depth, usually V-formed (ria coasts).
      2. Glacial coasts.

   B. Coasts created by land deposits:
      1. Fluvial deposition coasts:
         • Deltaic coasts.
         • Alluvial plains flooded by the sea level rise.
      2. Glacial deposition coasts:

- Moraines partially flooded by sea level rise.
- Drumlins partially flooded by sea level rise.

3. Aeolian deposition coasts (sand dunes advancing towards the sea), and

4. Marshy vegetation coasts.

C. Coasts occurring from volcanic activity:

1. Recent lava flows.

2. Volcanic eruptions or collapses (calderas).

D. Coasts created by tectonic activity:

1. Cliff coasts originated by the action of faults.

2. Coasts related to folding.

II. Secondary coasts mainly created by marine processes:

A. Coasts formed by sea erosion:

1. Straightened coastal cliffs due to erosion by the sea processes.

2. Sea cliffs with uneven form that have sustained the activity of sea processes.

B. Coasts formed by marine deposition:

1. Coasts created by the deposition of sediment and the formation of sand bars in river mouths.

2. Coasts that have advanced in the sea because of deposition.

3. Coasts characterised by the presence of sand bars expanding in the sea and sand spits along the coastline.

4. Coral reefs.

Sheppard's classification is useful but it has some disadvantages, as many classifications do. For example, erosion and deposition on a coast usually happen simultaneously. Some sections of the coastline are eroded and this material is deposited elsewhere forming sand bars. Consequently it becomes apparent that the characterisation that coast as a coast of deposition or erosion is tricky.

Classification according to Valentin (1952)

Valentin (1952) suggested a classification considering the advance or retreat of the coastline. He remarked that a coast's advance can be due to the emersion (emergence) of a coastal area, because of the lowering of the sea level and/or to the advance of the land towards the sea due to deposition. On the contrary, a coast's retreat may be caused by the coastal

Retreating beach in Corfu Island (Greece) (by A. Vassilopoulos, N. Evelpidou).

area's submergence, the sea level rising and the removal of material (erosion). This classification includes the following type of coasts:

I. Coasts where land advances towards the sea:

    1. Due to land emersion:
- Coasts of emerged sea floor.

    2. Due to depositions of organisms:
- Formed by depositions related to the coast's flora (mangrove coasts).
- Formed by depositions related to the coast's fauna (i.e. coral coasts).

    3. Due to deposition of inorganic substances:
- Marine deposition in environments of limited tide activity. This category involves lagoon-barrier coasts, and dune-ridge coasts.
- Marine deposition in environments with strong tide activity: This category involves tideflats and barrier island coasts.

II. Coasts where the land retreats from the sea:

    1. Sea level rise provokes glacial landforms' submersion:
- With indications of glacial erosion: fjord coasts.
- Without indications of glacial erosion: fjord coasts.
- Glacial deposition coasts.

    2. Due to submersion (submergence) of landforms that occurred from fluvial erosion:
- In areas presenting recent

Notches located on limestone formations in Samos Island (Greece). Their presence above the sea level is due to tectonic uplift (by C. Centeri).

folding structures: (embayed upland coasts).
- In areas presenting past folding structures: (ria coasts).
- In horizontal structures: (embayed plateau coasts).

    3. Due to sea erosion: cliffed coasts.

Valentin's classification, while it is partly descriptive, considers the conditions under which a coast was formed. The main advantage of this classification is that it also considers sea level changes in relation to the land, while at the same time it is based on observations indicating the land's advance or retreat.

Furthermore he has considered the indications of change in the development of coasts, which can be expressed as an interaction between

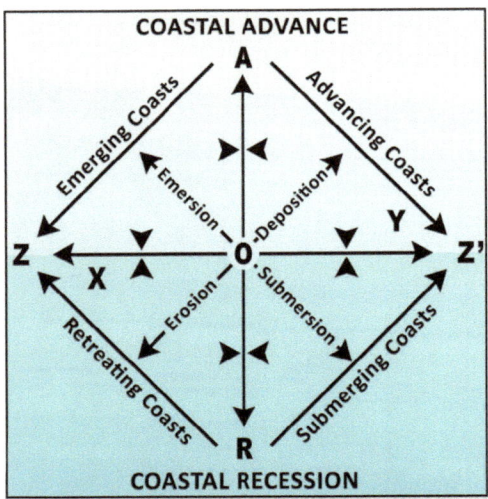

COASTAL ADVANCE

COASTAL RECESSION

Diagram of Valentin's classification representing coastline changes.

vertical (submersion, emersion) and horizontal (erosion, deposition) movements.

In the following diagram the ZOZ' line corresponds to coasts neither advancing nor being eroded, either because the land's emersion is counterbalanced by erosion (ZO line), or because the submergence is counterbalanced by deposition (OZ' line). The point O represents a completely stable coast where no changes of whatever nature happen. The greater changes are located in point A, where land's emersion in combination to the deposition, leads to a fast advance of the land towards the sea; and in point R where erosion in combination to the land's submergence, results to the coastline's fast retreat. It is obvious, that intense erosion can lead an emerging coast to retreat (point X), while fast deposition can lead a submerging coast to advance (point Y).

A category of coasts that need to be added to the above mentioned classifications, are coasts that have sustained alterations by the activity of humans on the coastal zone.

Coastal lagoon systems

Estuary systems and lagoon environments owe their creation to a combination of factors, such as for tectonism, that may uplift or depress an area, or erosion and deposition by wind, a river, or a glacial cover.

In the area where the river bed is broadened as it reaches the sea, estuary systems are created, with brackish water, whose salinity is regulated by tidal activity. These coastal lagoon systems are usually very productive since they have high concentration of nutrients provided to them by rivers.

The formation shallow coastal lagoons is possible, around the mouth of rivers or torrents. In these areas water is calm due to the protective role of the coastal sand bars. This leads to the formation of an ideal environment for the deposition of fluvial/torrential sediments. These longitudinal basins, known as *lagoons*, have complex features and particular sedimentary and hydrodynamic conditions. They may be isolated from the sea by sand bars or islets and may have subsaline or salty waters. In countries with warm, desert climate lagoons are transformed into lakes, as for example in the Sabkha environments. The opposite is happening in temperate zones or in low temperature areas.

In the cases where coastal basins are filled with sediments rich in organic material that originates from hydrophilic or aquatic vegetation, they are characterised by shallow and stagnant waters of low oxygenation, resulting in the formation of peats.

In low temperature conditions and in the absence of oxygen, dead vegetal material is accumulated in the form of peat in the lower bed of a coastal basin with shallow waters. When drainage is obstructed and the accumulated peat increases, a special environment is created in the estuary systems, called a *peat deposit*. Climatic conditions play a determinative part in the creation of peat deposits and they occur frequently in the more humid coastal areas of the Atlantic and in the alpine and continental areas.

Cultivated areas that have occurred from the drainage of previously aquatic areas often contain peat soils. Generally, estuary systems are the type of ecosystem with great biodiversity, since they are characterised by complex and particular hydrodynamic, ecological, biological and sedimentological conditions and a varied degree of isolation from the open sea.

Estuary systems are open systems, since they receive fresh water, from the rivers or the discharge of water tables, and salt water from the sea, thus directly interacting with their neighbouring ecosystems.

The equilibrium between the estuary ecosystems and their external environment is very fragile and any alteration in the exchange of material (sediment and water) and energy (by water movement, sediments and waves) can possibly lead to their degradation or even their disappearance.

Estuaries may be receiving material and water by fluvial processes, but also may have direct communication with the sea by a mutual contribution in material, depending on the natural conditions that prevail in each case.

Coastal basins of shallow water, like lagoons, can possibly provide sediment to the marine environment when the wave regime is capable of disturbing its bottom. This disturbance may cause the fluvial/torrential sediment to rise from the bottom. The finest-grained material is easily transported towards the marine environment over the sand bars. If the waves have enough height and energy and the sandbars are low, the fine-grained material is transported over them. In other cases the sediment can be transferred through natural channels of communication, which are named mouths or orifices.

Shallow estuary systems in general, and especially lagoons, are often eutrophic environments, due to the high supply of natural nutrients by the fluvial systems, but also due to the inflow of fertilisers from cultivated areas along the rivers and the coastal plains. As a result, the lagoons are characterised by a high content of organic material and nutrients, outmatching the system's autoconsumption capacity. The biogenic material surplus is either deposited as sediment on the basin's bottom, or is carried towards the marine environment. Seasonal and annual measurements have shown that there is generally high flow of the extra organic material from the lagoons towards the sea, while the opposite transfer (from the sea towards lagoons) is lower. It has been clearly shown that the eutrophic coastal basins provide the marine environment with carbon, phosphorous and nitrogen.

Active coastal dunes together with stabilized ones by the vegetation cover. Aberdeen (UK) (by A. Vassilopoulos, N. Evelpidou).

Evolution of estuary systems

The reasons for the disappearance of estuary systems are various and can be attributed to natural and non natural factors. A basic natural factor is their accretion by fluvial/torrential material. Another basic factor is the water level drop, which may be the result of human processes. Furthermore, when evaporation is intense and water supply is relatively low, the water level in estuary systems is lowered and after a long time it is drained. Depression of the water level may also occur after the opening of a lake's superficial drainage channel.

Human intervention can be determinative as far as water's disappearance from estuary systems is concerned. Irrigation and hydroelectric constructions result in the decrease of fluvial water supply to the coastal basin, which firstly results to the lowering of the basin's water level and afterwards to its drainage, provided that evaporation is higher than water supply. The decrease or cease of water supply implies a decrease or cease in material supply. The sand bands and sand islets formed and preserved by this material will be destroyed by the erosive activity of coastal currents and waves. This will result to the flooding of the coastal basin and will cause its disappearance.

Internal Circulation - Hydrodynamics

Coastal basins are characterised by particular hydrodynamic features and are ecosystems open to material and energy exchange. Their natural balance depends on the degree of influence local and global factors have on them.

Generally, internal circulation is taking place as a result of the combination, to a different degree each time, of factors such as the displacement , of the air and water mixture due to the tide, the water movement due to the density alteration on the vertical axis, the Earth's rotation (Coriolis Effect), and finally the formation of fronts and internal waves.

Tidal and wave activity generate currents within the coastal basin and directly influence the circulation of waters and sediments. An angular approach of the waves to the basin creates a current parallel to the coast, which contributes to the deposition and distribution of sediments along a sand band or a sand islet. The currents generated by low and high tide have a velocity that depends on the fluctuation of the tide, which is generally reduced in the inflow point of the basin where the fluvial system deposits. During intense rainfall conditions when fluvial supply is very high, the tidal currents have less influence.

Furthermore, strong winds and changes in barometrical pressure, lead to water movement and to a temporary circulation within the basin. Changes in temperature and salinity create smaller currents and weaker circulation.

Coastal sediments

In coastal basins, sediments may be distinguished in two main categories according to their origin:

- Clastic (originating from the land).
- Biogenic (originating from living organisms).

Clastic sediments originate from mechanical and chemical processes of land weathering and erosion, due to the activity of rivers and torrents. The biogenic sediments are directly produced by marine organisms and usually consist of bivalve organisms' shells or shell pieces. The species, from which the shells originate, depend on the physicochemical conditions of the environment.

The size and form of the sediment's components varies. Grains may appear individually or, due to the mixture of fluvial and marine water, they may appear in the form of floes (differential congelation) or pellets, due to the combination of bio-physico-chemical conditions. Usually the constitution of the coastal basins' sediment components is clay containing big pieces of bivalve shells.

Furthermore, the sediment's colour is usually very dark, even black, due to organic material decomposition. The processes of transportation and deposition are controlled by the grain size and mineralogy of the suspended sediments; those features are related to the sediments origin. Generally, clastic sediments are located at river mouths, while biochemical sediments appear in the central section of coastal basins.

Regarding the sediment size classification, the coarser material is accumulated on the island barrier and consists mainly of sand and pebbles. These components originate from the erosion and weathering of a coastal cliff and they have been transported by coastal currents. The fine components such as silt or clay accumulate in the basins' deepest parts, while the coarse ones accumulate in shallow parts.

main coastal landforms

ACTIVE CLIFF

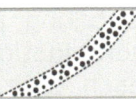

Topographic descent slope of high inclination, created by the sea's erosive activity; its form is defined by terrestrial dynamic processes.

Corfu-Greece (by A. Vassilopoulos, N. Evelpidou)

ALGAL REEFS

Organic rock structures in oceans and lakes characterised by a high concentration of carbonic salts. In areas where water is warm, they participate possibly in the creation of coral reefs, or in a smaller degree in the formation of unattached organic hummocks. The fossil algal platforms can be found on rocks of Precambrian age, and are known as stromatolites. Algal reefs are developed in areas where the conditions do not favour the development of corals.

Mihlemas Island-Australia (by K. Pavlopoulos)

BARRIERS

The barrier is a partially emerging ridge, consisting of sand or more thick-grained material and is developed either off-shore, or in shallow waters and almost parallel to the coastline. A barrier is usually cut by one or more straits where the tide enters, thus forming a chain of barriers – a succession of island barriers and peninsulas - or just narrow beaches. A partially emerged sandy tongue is called a barrier spit. The formation of the spits is connected to a coastline and a coastal current which does not follow the change of the coastline's direction when this is suddenly directed inland, but maintains its initial course. In that case the coastal transportation of sediments goes beyond the point of coastline shape change, resulting in their deposition and the formation of spits. These barriers are often described by terms such as sand banks, off-shore bars and barrier bars. Sandy reefs, sandy atolls etc, are considered to be similar formations. The barriers are narrow and are located near capes of variable coastlines, where a longitudinal development prevail. They initially begin with a simple deposition within the water, so that a kind of beach ridge is formed, usually covered by vegetation. Afterwards this is supplied with aeolian sand and an aeolian beach ridge occurs, which, if it develops to a hummock, then forms the front section of a dune or series of dune. Barriers are common and well developed along coasts with a low relief. This does not happen in areas where thick-grained sediment is rare or totally absent, where the wave energy is low or in areas where the tidal width exceeds 30m. As for barrier coastlines, they are very well lined-up where the energy of waves is high and poorly lined-up where the energy of waves is moderate. Barriers are transformed into a series of small islands (like a fragmented barrier) in areas where tidal width and sand supply approach critical values. Consequently, the barrier islands are zoned deposits which lie parallel to the coastline and are separated from it by shallow bays and lagoons. Usually, their creation is related to the combination of low tidal and wave activity.

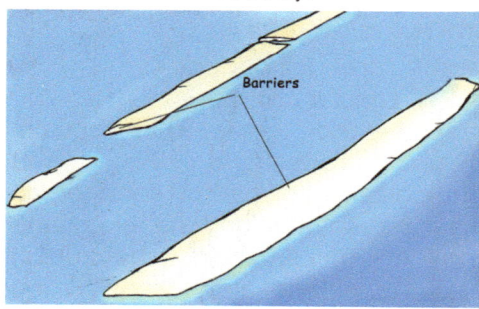

BEACH CUSPS

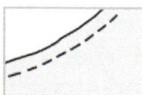

Formed parallel to the waves' movement, the beach cusps are developed as a series of crescent shaped sand concentrations on or slightly above the coastline and they consist of crests with troughs between them. The crests have an almost triangular shape with rounded tops which expand in the water; their height is usually low. However, there are cases where the crest height exceeds 1m. The distance between the crests varies from a few centimetres up to a few tens of meters. Beach cusps are developed early in the year, during the change between winter and summer, when wave energy

is decreased. Their formation is also related to the currents which predominate in the fracture zone; they are found on sediments that vary from big roundstones and pebbles to fine-grained calcic and siliceous sand. One of the important geomorphological characteristics of these landforms is the fact that the thick-grained material is accumulated in the crests, while the more fine–grained material is accumulated in the troughs.

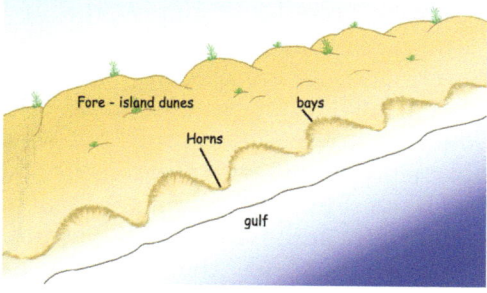

BEACHROCKS

The term Beachrock is used for the description of specific rock formations of the coastal zone that consist of sand and coarse material, such as roundstones, pebbles and etc., with calcite or aragonite as their cement material. Although there are many reports on recently generated beachrocks in coastlines of cold or temperate climatic conditions, it seems that these landforms mostly develop in tropical and subtropical areas. Many extensive studies by a great number of scientists, in many different places have taken place on beachrocks and there are many of theories related to issues such as their age, their way of formation, the areas where they are created and the kind of their adhesive material. Beachrocks are formed even today, so the study

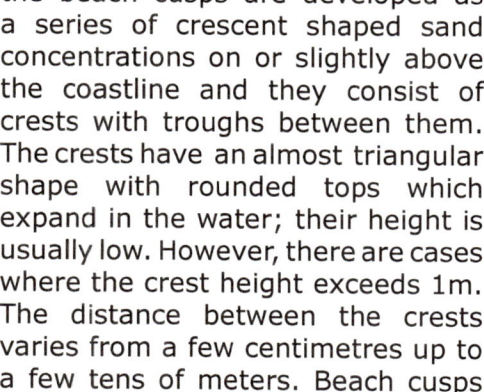

of their development is possible, as is the study of the conditions of their diagenesis. Nevertheless, most of the studies have been based on already compacted and old beachrocks. Opinions differ regarding their place of origin since they are visible only long after the completion of their cohesion process. The majority of scientists agrees that their cohesion process happens undersea, maybe within or near the upper section of the water saturated coastal zone. According to this theory, it becomes clear that beachrocks appear only in the areas, where, after the completion of the cohesion process, the sand that was not incorporated with the beachrock has been drawn away during beach retreat. In areas where the retreat is of small scale, the beachrocks appear as a flat platforms of various widths (from a few meters to tens of meters) in front of the beach, and stretch along the beach for tens to hundreds of meters. However, in places where significant retreat of the beach has happened, the beachrock shores remain out at sea as reef ridges. In the cases where coral reefs exist, beachrocks that appear out at sea, in the form of reef ridges can be falsely considered to be coral reefs, especially when they are covered with seaweed and other sea organisms. In such areas, their sand may possibly originate totally from calcite organic residues. This has lead many researchers to the conclusion that all the beachrocks are calcitic, but Boekschoten (1962) and others have studied some beachrocks which contain little clastic material of carbonate composition. Contemporary researchers agree that the beachrock shores have

generally a similar composition to the materials that predominate in the beach of the area where they are formed.

Kineta-Greece (by K. Pavlopoulos)

BERM

Berms are formations which consist of round pebbles/sand zones that usually present concentric form and lithologically consist of sand of all grain sizes, of round pebbles, and other microfragments of the coastal area. An oblong arched sand zone is formed along the coastline by the deposition of the aforementioned sediments on the highest section of the coast to be reached by waves. The creation, the location and the number of the concentric sandy coastal zones depend on the activity of waves. Every time the energy of waves is changes, a new berm is created, while possibly an older one is destroyed. In this way, the formation of more berms than one, shows the existence of different wave and transportation capacities in the area. In particular, the highest berm (the one which is located at the highest point of the coast) where the more thick-grained material is concentrated, represents a high energy activity of the waves, while the lowest, where concentration of more fine-grained material takes place, represents low activity.

Rouen-France (by A. Vassilopoulos, N. Evelpidou)

COASTAL CAVE

A cavity in the coastal rocks of an area, which has been created by the erosive activity of waves. Carbonate rocks are mort susceptible to the creation of caves. Some caves expand greatly, penetrate small capes and form impressive arches. The collapse of the roofs of some caves brings detached pieces in front of the rocky beaches.

Zakynthos-Greece (by S. Liakopoulos)

COASTAL PLATFORMS

Flat rock benches which are created by sea erosion between the highest and the lowest sea level. Biological, chemical and mechanical processes, are considered to be the most important weathering factors. They play a primary part in platform formation, while the wave energy, which is the main factor of transportation of the erosion products, plays a secondary part. Coastal platforms can be also created by the protracted erosion of the sea notches of the slope's front. In this case the platform's form and development is defined by the slope's lithology and stratigraphy. Finally, some coastal platforms are related to eustatic movements, while the presence of a particular type of terrace is usually attributed to the change of the type of incident waves.

Aigina-Greece (by A. Vassilopoulos, N. Evelpidou)

COASTAL SAND DUNES

Landform which is created when material transferred by the wind (usually sand) meets some obstacle during its transfer. (i.e. vegetation, branches, ditches, protrusions, etc). The morphological features of dunes are related to the quantity of sand which can be transported by the wind and by the cycles of deposition-erosion. The dimensions of the dunes vary and their diameters range from a few meters to many kilometres. Their height varies from 1 or 2 m up to 20-30 m.

In particular, dunes that are created by winds of constant direction, have slopes unequally tilted. More specifically, the windward side has a low inclination which ranges

89

between 5-12°, while the lee side has a higher inclination which can reach up to 20° to 30°.

Regarding their shape, it also has high variety, from very simple to composite forms. Thus, there are crescent, longitudinal, matterhorn dunes and also dunes which are created by very complicated combinations.

Dunes can develop in every subaerial environment where loose material, of sand grain size, are exposed to the wind's activity and can easily migrate and accumulate in large masses. Every obstacle on the ground, such as protrusions, ditches, the presence of vegetation etc, contribute to this accumulation. Furthermore, the presence of humidity may stabilize the sand accumulation, beginning the creation of a dune.

It is a fact that the coastal dunes cover a very small area compared to the vast areas of dunes which are generated in the deserts. Coastal dunes are generated where a broad sandy coast exists; this is usually characterised by a great tidal width and rarely by rocky and steep coasts. Many sand dunes originate from older geological eras when sea level was much lower than it is today. The sand is transported towards the land in a bouncing way, during which the movement of grains is accelerated after they are raised by the wind. The wind speed is slower close to the ground and faster as we rise, since the ground's friction is reduced. When a cloud of sand passes over an uneven surface, its speed is reduced due to the loss of energy. This results to the deposition of transported material and leads in the creation of a new dune.

The development and stabilisation of a dune depend on vegetation which entraps the sand and reduces the activity of aeolian energy. After the development and stabilisation of the dune, its migration towards the land's interior may possibly follow, as its seaward side might be eroded depositing its material along the shoreline. When the Aeolian energy exceeds the cohesion of the sand layer or the resistance of vegetation, large parts of the dune can possibly be detached.

Aberdeen-UK (by A. Vassilopoulos, N. Evelpidou)

COASTAL SEDIMENTS

The coastline consists of sediments of different classification (sand, gravel, silt, blocks, etc) which reflect the dynamics that prevail in the coastal environment. When the material of a beach consists of homogeneous and fine sediments, it is related to an environment of low and constant energy; however, if the material is heterogeneous, it is related to an environment with a highly variable dynamic. The presence of roundstones possibly indicates the beach's supply by a torrent. Such material is often found in average energy environments. Material like sand or clay could have originated from a deep submarine basin, since it can be transported over great distances by sea waves and currents, or it could have originated from river mouths (i.e. deltas).

Samos-Greece (by C. Centeri)

slope, while relatively high dynamic pressures occur when waves of a vertical front plunge on the cliff with their top and their trough simultaneously.

Sporades-Greece (by A. Vassilopoulos, N. Evelpidou)

CUSPATE COASTAL SLOPE

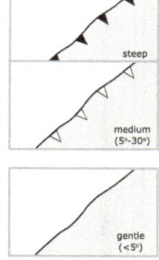

Steep coastal slopes which occur as a result of marine erosion. The slopes' resistance to erosion is a function of the wave energy and of the cohesion of the rocks which constitute the slopes. One could mention that the erosional role of sea waves is double, as it does not only erode the base of the slope's front but also removes the weathering products from its base. When the weathering products are accumulated at the slope's front base, its retreat will possibly slow down or stop, since they protect it from the incident waves. At a slope front two types of pressures are exerted; the one is related to the weight of the incident sea mass (static pressure) and the other one depends on the wave type, which is a function of the waves' dimension and of the inclination of the slope front (dynamic pressure). We have the lowest dynamic pressures when the waves are fully reflected or when they break before they reach the

CUSPATE FORELANDS

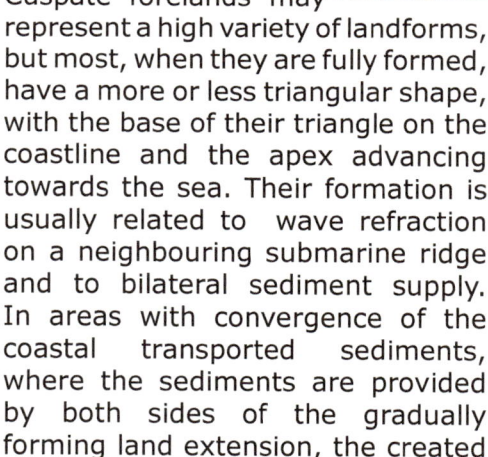

Cuspate forelands may represent a high variety of landforms, but most, when they are fully formed, have a more or less triangular shape, with the base of their triangle on the coastline and the apex advancing towards the sea. Their formation is usually related to wave refraction on a neighbouring submarine ridge and to bilateral sediment supply. In areas with convergence of the coastal transported sediments, where the sediments are provided by both sides of the gradually forming land extension, the created cuspate forelands have more intense seaward development.

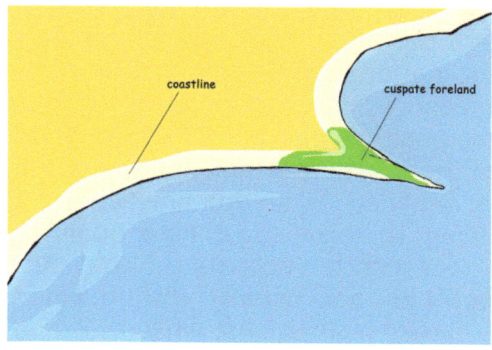

91

FLOODED FLUVIAL VALLEYS - RIA BEACHES

These belong to primary coastlines, and owe their formation to sea impact on a landform created by terrestrial factors. They are identified by the shallow waters of the valleys' sunken rivermouths which cut the land serrately, presenting a rich and complex horizontal dissection. When not interrupted by some natural barrier, their axes dip towards the sea. The characteristic types of Ria coasts are the dendritic type, the shape of which resembles an oak leaf and is due to the fluvial erosion of horizontal layers of homogeneous material, and the network type which is due to the fluvial erosion of tilted layers of different hardness.

Kefalonia-Greece (by K. Pavlopoulos)

INCLINATION OF THE COASTLINE

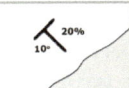

The distance from the beach to the relatively flat area which comes after the beach front, is a parameter which influences the inclination of the coastline, which is measured in degrees or %. If this distance is small, then the inclination of the coastline is high and the beach is qualified as steep; however, if it is big, then the coastline is qualified as of gentle inclination. The rock beaches are usually distinguished by steep inclinations, while the sand beaches by more gentle inclinations.

Aigina-Greece (by A. Vassilopoulos, N. Evelpidou)

LONG SHORE CURRENT

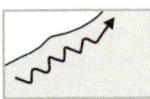

A very powerful coastal current, due to which sediment is transported along the shore.

Kavala-Greece (by Keramoti municipality)

LONGSHORE BARS AND BEACH RIDGES

Landforms which look like sand rumples, but are bigger and have lower normal gradation. They are usually created in shallow epicontinental environments or on the shelf borders, by the activity of waves and coastal currents, and are found individually or in groups. The longshore or sand bars are classified as longitudinal or transversal in relation to the predominant coastal current or the coastline.

Longitudinal longshore bars, are found in river mouths, in funnel-shaped river bays, in straits and also in creeks, where the tidal effect is observed. As far as transversal ridges are concerned, the crescent-shaped longshore bars, which are found in river mouths and in tidal creek channels, are typical examples.

Longshore bars are formed of sand that moves parallel to the coastline. Particularly during low tide, they may be uncovered and exposed to atmospheric activity. Often several longshore bars are formed, in one or more series, which are arrayed parallel to each other, and at different depths in relation to the sea surface. The term beach ridges is used to describe a series of longitudinal and parallel ridges, consisting mainly of sand, shells and roundstones , varying in width from a few centimetres to a few meters andat intervals of 25-500 m. They are usually located behind the contemporary beach. The ones that are found in deltaic environments, appear concentrated on a muddy substratum and are known by the term cheniers. Every ridge indicates the position of the paleo-coastline. Usually, the beach ridges are, as far as their construction is concerned, the natural evolution of some coastal landforms, such as the longshore bars or the cuspate forelands. Many researchers consider that most of the beach ridges which can be found today were formed after the stabilisation of the sea level at the current levels, that is in the last 4.500-6.000 years approximately, a period which coincides with the Flandrian transgression. The sandy beach ridges may possibly contain also a percentage of transferred aeolian material. This however rarely occurs in places where successive ridges have been developed within a very short time

period or in warm and humid areas, where the development of dunes is not possible. Since the sandy beach ridges are directly connected with fossil beaches, they sometimes contain significant accumulations of heavy minerals of great financial value.

Kavala-Greece (by Keramoti municipality)

MARINE TERRACES

The continuous wave activity in the coastal zone generates a typical coast profile, which consists of a sea slope and a submarine terrace. The sea slope begins as a low form, whose height is increased towards the inland, while it remains as a submarine platform in its base. The material which comes from the slope's weathering is transported by the bottom currents and deposited off the edge of the rocky terrace, resulting to the creation of a terrace generated strictly by the waves' activity. Along some beaches, the coastal currents are so powerful, that the sediment coming from inland erosion is carried away, so that the only remaining landform is the platform generated by sea erosion. If the sea level remains stable for a long time or if the rise of the sea level happens extremely slowly, the sea cliff will be quite far inland, while the terrace resulting from the wave activity will be quite expanded. If the evolution period is shorter, this wave cut terrace will be less wide. The changes in sea water volume are not the only causes that contribute to coastline changes. If the sea level rises and reaches a new level where it can remain constant for quite a time period, a new sea slope and a new platform will be created. Thus, every period of sea level stabilisation is followed by the creation of a sea slope and a platform. If the sea level is lowered, its former levels will become apparent through a succession of terraces. The topographically higher terrace, having sustained weathering and erosion for a longer time period, tends to become indistinct and is usually represented by an increase of inclinations at the locations where the older sea slopes existed. The terraces that were formed when the sea level was on lower levels than it is today, have been flooded when the sea reached its current level. As the rising sea has covered these terraces, they will have suffered considerable erosion by waves and will have been partially covered with material.

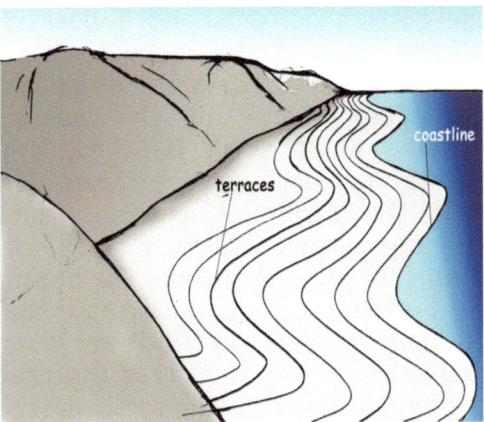

MARMITE

Round ditch created by the turbulent movement of the roundstones which are transported by waves or by turbulent currents.

Limbe-Cameroun (by K. Pavlopoulos)

NOTCH

Formations located on rocky coasts. They are located in places where the sea surface meets the land and are created due to processes of friction, solution or biological factors. Since during the last years sea level is rising, their presence above sea level indicates tectonically active areas, where the land is rising. Therefore, by studying the sea fauna in these notches, we collect characteristic data for periods of constant rise , for the rising rate and for the earthquake risk of the studied area.

Samos-Greece (by C. Centeri)

RETREATING BEACH

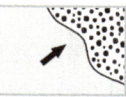

The exact opposite of the advancing beach; if the beach consists of loose sediments, the erosion factor clearly depends on the dynamic of waves and on their ability to transport material. During the beach's retreat entire zones of beach ridges or even dunes can move. As in the case of the beach's advance, during the retreat, the basic formation factors are time, energy, sediment supply, the change of sea level and the development of vegetation. Time guarantees a complete dynamic counterbalance after a change in one of the factors. Energy, in the form of sea currents, is increased during intense weather conditions and accelerates the retreat. The decrease of sediment supply in areas where tidal currents exist, also leads to the aggravation of the retreat. The existence of vegetation in the dunes decreases the erosion rate during a temporary retreat.

Kefalonia-Greece (by A. Vassilopoulos, N. Evelpidou)

ROCK MUSHROOM

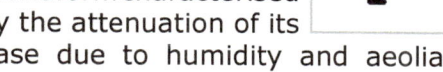

A landform characterised by the attenuation of its base due to humidity and aeolian erosion.

Kefalonia-Greece (by A. Vassilopoulos, N. Evelpidou)

SAND BEACH

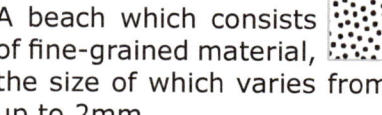

A beach which consists of fine-grained material, the size of which varies from 50µm up to 2mm.

Los Roques National Park-Venezuela (by C. Centeri)

SEA ARCH

The sea arch is a natural opening at the front of a coastal slope, and is created due to marine processes of erosion. Arches are developed in areas with a lithological and tectonic status which allows the creation of coastal caves by wave activity. Their creation is similar to that of coastal caves. Two caves that are created on both sides of a cape may meet after a long time span, first forming a tunnel, and finally an arch as the erosion progresses. The central part of the arch's roof, is known as the "keystone" and it supports the entire structure. The architectonic structure of an arch reflects the hosting lithology. The arch's shape may be arcuate or rectangular, submarine or not and the height of their opening may reach up to tens of meters above the basic level. Sea arches are considered as ephemeral landforms of differential erosion and exist only for a few decades or centuries.

Samos-Greece (by A. Vassilopoulos, N. Evelpidou)

STACK

Rocks of pyramidal shape that protrude in the sea. They are created when the slope retreats, leaving erosion residues at the sea. The sides of stacks are generally steep and vertical, a fact which indicates that the erosion has taken place at wave height and not below the sea surface. The term stack comes from the word stakkur, in the Scandinavian dialect of the Faeroe islands, where the particular landforms are very often found in front of high, rocky beaches. Often, in the foreign bibliography, the terms pillars, chimney, rock column, skerries, needles etc., are used.

Olympic National Park-USA (by C. Centeri)

TOMBOLO

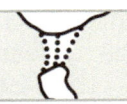

A Tombolo is a landform which is formed when a cuspate foreland connects the coastline with an islet, rocky or sandy. The term Tombolo initially originated from Italy and was referring to one or more sandy tongue-shaped formations which were connected to the land. It is a quite usual landform along flooded coastlines that are in their youth or at the beginning of their maturity. In the areas where a double Tombolo is formed, a lagoon between the two landforms is created, which is gradually filled with material and thus a broad, flat mound is formed. Gibraltar is a typical example of a double Tombolo. The world's greater Tombolo is considered to be the one that formerly connected Ceylon (Sri Lanca) to India along the Palk Strait, which is known by the name Adams Bridge. The particular landform was destroyed during a small scale change of the sea level which has taken place many thousands of years ago and which remains today as a series of islets.

Naxos-Greece (by A. Vassilopoulos, N. Evelpidou)

LAGOON LANDFORMS

LAGOON

A basin of longitudinal shape which is located

along the coastline, very close to it, and is separated from the sea by island barriers. Usually it is developed diagonally to the estuary of one or more torrents; the calm waters behind the island barrier are an ideal environment for the deposition of fluvial/torrential sediments.

of sediment and water between the two environments.

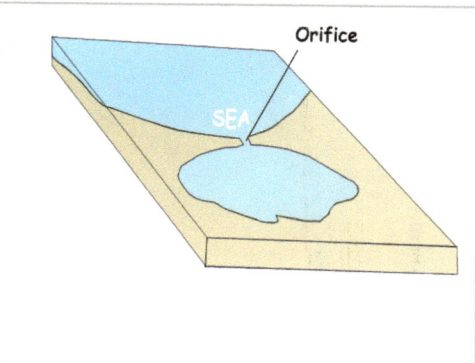

Naxos-Greece (by A. Vassilopoulos, N. Evelpidou)

MARSH

An area of stagnant waters of little depth, characterised by aquatic or hydrophilic vegetation. The water and the sediments of these areas are usually very dark coloured, even black in some cases. This is due to the presence of much decomposing organic material.

Samos-Greece (by A. Vassilopoulos, N. Evelpidou)

MOUTH OR ORIFICE

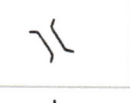

A natural opening of the lagoon towards the sea, which facilitates the exchange

Osterseen lake (Upper Bavaria-Germany) (by O. Bender)

Chapter 4

lacustrine environments

lacustrine processes

Lakes-Introduction

In the second half of the 20th century, the famous Swedish limnologist Forel has defined lake as a mass of stagnant water located in a trough of the ground and has no direct contact with the sea. It can be geologically considered as a temporary water mass, appearing or disappearing within a short time period.

Today, lakes are located everywhere on our planet. However, they are more frequent in higher geographic latitudes, and in mountain areas. They are principally common in glacial and periglacial areas, especially where the alteration, from glacial geoenvironments into more warm and humid ones, was quite recent, and also along rivers with low inclinations and broad valleys, where they connect to other branches.

Lake water can be either fresh or salty. This depends mainly on the prevailing climatic conditions in the area. Lake water originates directly from precipitates, from water springs, from runoff water, or even from the sea.

Lakes, although they are open systems regarding material and energy, they are examined and analysed as independent systems and are characterised by special physical, chemical and biological parameters linked to their degree of isolation and the geographic position of each lake.

History of the existence of lakes

All lakes have limited life duration and in general follow a disappearance course. In humid areas their disappearance begins after the erosion of their barrier, the outflow of its water and the deposition of sediments and organic material on deltas or on basis deposits. During their short history, their chemical composition

Salty lake (Marsh) at Samos Island (Greece) (by C. Centeri).

A group of lakes which originates from ice melting. These lakes are characterised by an extraordinary diversity of hydrological and chemical properties. Osterseen (Upper Bavaria, Germany) (by O. Bender).

does not significant change. In arid areas, lakes disappear due to higher evaporation and to deposition of material transferred by wind and water. Due to evaporation, many lakes in arid areas are gradually made saline, even if the initial lake was a fresh water lake.

Classification of lakes

A lake can be formed by one or more factors. Various specialists have classified lakes in different ways, for example a classification based on the conditions that may have possibly created the basins, and have termed them creative, destructive or retarding. Other scientists have classified lakes on the basis that they are formed within troughs consisting of bedrock, in basins formed by natural or artificial barriers, or are organic lakes. Both systems can possibly be criticised because they exclude natural, territorial groupings. The limnologists dealing with one group of lakes should consider the conditions that lead to their formation.

Hutchison, taking these positions, based a classification on the origin of lakes, which is presented below, simplified.

Lakes of tectonic origin

In this case the basin may have been formed in one of the following ways:

I. By gentle movements of the crust. This category includes:(i) Residual sea basins which have been isolated due to continental movements e.g. the Caspian Sea. (ii) Lakes created because of sea level rise i.e. Lake Okeechobee, Florida. (iii) Lakes located in areas with mild inclination that may eventually lead to the runoff inversion, e.g. Lake Kioga, Eastern Africa. (iv) Lakes having

101

Maclu lakes - Jura (by O. Bender)

a central basin, formed because of the mild elevation of the area's borders, e.g. Lake Victoria.

II. By the elevation of peneplains during orogenetic movements. The created basins appear between the mountains and can lead to the formation of lakes. In some cases local faulting may define the border of the lake i.e. Lake Titicaca, Andes.

III. Due to the folding of geological formations.

IV. Due to faulting. They are an important category of lakes. Many of the world's largest lakes belong to this category i.e. Lake Baikal.

Basins related to volcanic activity

I. Lakes formed within modified or partially modified craters.

II. Lakes formed in calderas.

III. Lakes in modified calderas where local faults play an important role.

IV. Lakes in collapsed lava flows.

V. Lakes formed within barriers which originate from lava, volcanic silt or volcanoes.

Lakes formed by soil or rock subsidence

The lakes that belong to this category have usually very short existence.

Lakes formed by glacial activity

The lakes formed by glacial factors constitute a special category, since they were formed during a very short period of the Earth's history. During the Pleistocene, glaciation has created lakes more than any other landform. Some typical categories of lakes that originated from glacial activity are:

I. Lakes behind barriers of ice.

II. Lakes in glacial rock basins: (i) The cirque or corrie lakes form almost at the snow border in glacial valleys. (ii) Lakes formed in basins consisting of bedrock, behind the snow border, due to glacial erosion. (iii) Lakes formed by continental ice. (iv) Lake-like basins formed by glacial dissolution. The lakes of this category are usually small.

III. Lakes as a result of glacial deposits. The glacial moraines many times constitute the barriers for the formation of lakes.

IV. Drift lakes or kettle lakes. Lakes formed when water fills the small soil depressions (kettles) that have been formed by the melting of ice blocks buried in the sediments of a glacial outwash plain. It is a common lake category, but usually of small size.

Lakes formed by the dissolution of rocks

I. The dissolution of limestone by water results to the creation of karstic basins of almost circular shape. These basins (dolines, uvalas or poljes), are drained through a series of sinkholes or natural drainage pipes. When sediments or other obstacles set by tectonic activity etc. block the drainage paths, these basins can be filled with water and create karstic lakes, such as the doline lakes.

II. Lakes can be created by the dissolution of rocks in basins delimited by tectonic characteristics (e.g. faults).

III. Lakes created by soil depressions after the natural dissolution of salts in the underground soil layers.

Lakes formed by fluvial activity

Some reasons that may lead to this type of lakes are:

I. Fluvial erosion.

II. Fluvial deposition: (i) alluvial cones and the deltas that separate the existent lakes, (ii) residues of a principal river that block an area, (iii) basins formed by abandoned channels in flood fields.

Lagoons

Usually lagoons are created behind barriers, spits and tombolos. Their formation is favoured by sea level rise, which floods the rivermouth and feeds the lagoon with seawater. Later, when sea level has lowered, this connection is cut and the sand barriers delimiting the lagoon are stabilised.

Lakes created by the activity of wind

This type of lakes can be created in basins formed in one of the following ways:

I. Basins barred by material drifted by the wind.

II. Basins formed between dunes.

III. Basins created due to the removal of material by the wind.

Lakes formed by the accumulation of organic material

This category includes lakes created in basins that have been isolated due to the formation of natural barriers by the dense development and accumulation of organic material i.e. plants. Washington Island Lake in the Central Pacific Ocean is a coral atoll basin located above sea level and belongs to this category.

Lakes created by a meteorite impact

Lakes created in the crater formed by meteorite. Usually the presence of water in the crater is related to the accumulation of runoff water. If the crater is deep enough to meet the aquifer level, the water can discharge from the water table and form the lake.

Lakes of anthropogenic origin

This category includes all the flood basins created by reservoirs and dams.

Lake water: Compostion, Movements and Properties

1. Composition

The quantity of salts in lake water varies significantly. For example the Great Salt Lake, Utah U.S.A. contains 238,12gr/lt, while the lake of Geneva contains only 0,1775gr/lt. The quantities of salts dissolved in lake water are a result of the lake's initial composition, of the salts that come into it and of the degree of evaporation.

2. Water movement in lake systems

The water movement in a lake is usually turbiditic. Everywhere within a lake environment, water movement towards any direction is taking place. The turbiditic movement allows the transportation of material and heat towards every direction and increases the apparent viscosity of water. We may also examine the significance of currents

A small moor lake formed after glacial processes. This lake has recently been filled with organic biomass. Seewaldsee (Austria) (by O. Bender).

determined by winds. The prevailing currents are: a) those due to the movement of inflowing and outflowing water, b) tidal currents, c) density currents generated by the difference of temperature and load beneath the main water mass, d) those generated by wind; this causes small disturbances and wrinkles on the water surface, and also generates an inversion current, where water flows in the opposite direction. If, during flow, the energy of the water is lost, a new flow begins at the terminal point of the old flow. In general this generates a periodic movement or periodic oscillation of the water. Furthermore, the periodic oscillation of lake water can also derive from uneven atmospheric pressure over a lake. The oscillation period depends on the shape of the lake.

3. Temperature

Lake water temperature varies depending on the season and the position within the lake. The factors controlling temperature are the isolation of the lake, the atmospheric temperature, the inflow by rivers and the precipitates. Temperature differences within the lake mass can lead to the lake's water stratification, which can also be generated by variations of salinity or of quantity of dissolved sediment. Lakes lacking complete water circulation are divided in a superior area of warm water circulation, which is quite turbulent and is called epilimnion, and a deeper area relatively undisturbed called hypolimnion. The two areas are separated by a level of rapid temperature change called a thermocline.

Stratification can be destroyed when surface water becomes colder, giving it a higher density than that of deeper water. When the cold surface layer is sinking, water layers invert. Inversions of that kind tend to be seasonal. In lakes where the temperature of surface water does not fall below 4°C (the temperature where fresh water is densest), an inversion appears during autumn. In lakes where the temperature of superficial water falls under 4°C, there is the possibility of two inversions per year.

Sedimentation in lake environments

Freshwater and salty lakes are characterised by different sedimentary deposits. In many cases, in salty lakes, the existent deposits due to evaporation are interrupted by clastic sediments entering the lake in flood periods. The composition of evaporitic sediments may also vary. Sediments of organic origin are rare in these lakes, because the conditions do not favour the development of life.

The sediments in freshwater lakes vary significantly and depend on a multitude of factors. These include the origin of the lake's basin, the character of the rocks and the ground surrounding the lake and in its drainage area, the size and depth of the basin, the expanse of shallow water near the coastline, the relief, the percentage and type of the drainage basin's vegetation cover, the climatic conditions and the organisms living in the lake.

In freshwater lakes, recent deposits often contain high percentage of organic material. Wasmund (1930) has suggested some terms to name the depositions of organic material involved in freshwater lake sediments. According to Wasmund, the residues of animals are called Forna. Forna are different than the vegetal and animal residues of colloid size, which are called afja. The Gyttja is a deposition form of organic material under acidic conditions. These deposits rich in organic material may rapidly increase in the stagnant water of some lakes. They can also be dated through microscopical observation of the vegetal residues trapped within the sediment. These residues can also act as an indicator of climatic changes during their deposition in the lake. The dating of organic sediments can be carried out with great precision using the ^{14}C- dating method.

In a lake, the sediments deposited are rapidly altered by bacterial activity. All new sediments are affected by biogenic processes induced by macro- and micro-organisms.

In lakes that receive water from the melting of the glacial cover or have done so in the past, the varve sediments are of particular interest. The varves are the annual products of a sedimentation cycle. The rapid melting of a glacial cover during spring or summer implies the release of great quantities of water. This leads to relatively course-grained material reaching the lake, while the more fine-grained material is deposited during winter or during glaciation, when inflows are reduced. The layer formed during the winter is clearly separated from the one formed during the summer. The two layers reflect the annual deposition and are useful for the determination of a period before and after the glaciation.

The formation of Vouliagmeni Lake is attributed to the collapse of a cave. South Attica (Greece) (by E. Efraimiadou).

main lacustrine landforms

CRATER LAKE

A lake that has been formed within an inactive volcanic crater or in a crater created by a meteorite impact.

Laghi di monticchio-Italy
(by O. Bender)

LAKE DOLINE

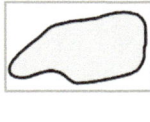

Usually the bottom of dolines is covered by silt and clay deposition material which prevent the infiltration of water resulting in the formation of doline lakes.

Triglav National Park-Slovenia
(by O. Bender)

PERMANENT LAKE

Lake which keeps its water throughout the whole year.

Lago di Gusana-Italy (by O. Bender)

TEMPORARY LAKE

A lake which can appear temporarily for short or long time periods. Its extinction is due to lowering of its water levels, either because of the opening of a superficial outlet channel , or because of the widening of its sub-water leakage channels. Climatic conditions such as intense evaporation for long periods, are also of great importance.

Donana National Park-Hungary
(by C. Centeri)

SALT LAKE

Inflowing lake, whose water has a salt content of less than 5% (i.e. NaCl, Na_2CO_3, $CaCO_3$, $CaSO_4$, etc). When salty lakes are located near to marine environments, they are usually sea inlets that have been separated from the sea by coastal sand barriers or by protrusions generated by tectonic activity. Sabkha lakes belong in this category. Sabkha is an arabic term which is used for flat basins, generally situated a small distance from the marine environment and covered by a layer of salt. They are typical forms for the North African and Arabian coastal areas. Sabkhas can be periodically covered by water which comes from atmospheric precipitates or tides in the coastal areas. Many of these coastal basins are situated slightly above sea level and are the results of small eustatic movements of the Middle Holocene.

Salinas Grandes del Noroeste-Jujuy (Argentina) (by S. Kanitscheider)

Clavell glacier - Canada (by C. Centeri)

Chapter 5

glacial environments

glacial processes

Creation and expansion of glaciers

The genesis of a glacier is a gradual natural process happening at various rates which depend on the prevailing climatic conditions. The areas covered by snow throughout the year can be defined as areas of eternal snow. When the snow falls in regions of temperate climate and low elevation it stays frozen for a few weeks and then melts. In contrast, in polar areas the snow constantly accumulates with its density ranging from 0.1 Kg/m³ to 0.9 Kg/m³. The accumulation and subsequent compaction of snow creates the glacier which can now move into lower areas. This is due to the gravity effect and seasonal temperature changes. The formation of glaciers may depend either on low temperatures (for the accumulation of snow) or on the preservation of snow (requiring low evaporation and snow loss during the summer). A permanent snow line appears in

Glacier within the ablation zone. Ice leaves the glacier system by melting and evaporation. Clavell glacier (Canada) (by C. Centeri).

all continents except Australia. Its existence depends on altitude and geographic latitude. For example, in polar areas snow line may be found at the sea level, in Greenland at 610 meters, in the Alps at 2,740 meters and in Africa at 5,480 meters.

The constant accumulation of snow in areas with steep slopes results to the generation of avalanches which contribute significantly to the formation of snow layers of great thickness in places of lower altitude and milder relief. Then, solar radiation melts the surface layer of the snow mass and the produced water intrudes into the rest of the frozen mass. There, the meltwater refreezes at a very fast rate creating a pile of crystal grains. In this way granular ice is formed. If the local climatic conditions favour the last process, it is repeated and granular ice becomes more compact, in this way getting an effervescent structure due to the air bubbles have been trapped inside the ice mass. The constant increase of the compressive tensions causes ejection of the air and further compaction of the ice mass leading to the glacier formation.

Then, under the influence of gravity the glacier starts to move. Its flow is achieved through the process of ice refreezing. During glacier movement, the ice mass breaks into several pieces which, through the refreezing process, are rejoined together, giving the glacier the appearance of a homogeneous and plastic mass. The sliding of the glacier takes place on its base and margins, whilst its internal body is characterised by hearing forces. The displacement speed of the glacier

Piedmont glacier and debris flow. Glacier National Park (Canada) (by C. Centeri).

shows a linear trend and has its maximum value in the central part of its main body. Furthermore, there will be an extensional or compressive displacement (deceleration) depending on the topography.

An important factor that controls the way in which the glacier moves is the thermal condition of the glacier base. It has been observed that several glaciers demonstrate base temperatures lower than the fusion point at a given pressure, in contrast with others which exhibit higher temperatures resulting to the melting of their base. Therefore, glaciers characterised by a cold base may be mobilised by internal deformations, while those characterised by a thawed base may move by sliding and internal deformations.

The movement of the glaciers leads to their expansion beyond the areas of eternal snow. The speed of their displacement ranges from a few tens of meters per day up to several kilometres per year and depends mainly on the ground slope and the variation in the size of their mass. During winter, when the mass of the glacier increases, a consequent increase in the glacier displacement speed is observed, whilst during summer, a reduction in the glacier mass leads to a decrease of its displacement speed.

Glacial weathering and erosion

The detachment of large pieces of rock mass from a valley due to the glacier movement is called frost weathering. The water produced by the melting of the glacier usually penetrates the rock fissures and, while refreezing, may cause mechanical detachment of large rock

113

masses that may be carried for a long distance by the glacier. The pressure fluctuations while glacier moves can lead to dynamic erosion effects due to energy release by the pressure variation. Comparative studies have shown that the intensity of erosion on different relief types covered by ice varies considerably and always obtains its maximum value on the leeward side of the relief. The severe erosion of the relief due the detachment of rock fragments by the glacier is usually known as glacial abrasion. Necessary prerequisites for the existence of abrasion is the carriage of hard rock residues by the glacier base, the sliding process, and satisfactory ice mass thickness for the development of high pressures on the base of the glacier.

Sometimes, in mountainous regions the bottom areas of the valleys may be covered by glacial debris. This debris is derived from the excavation of the bedrock by the glacier. Excavation is a significant erosion factor and produces large fragments of rocks. In contrast, abrasion smoothes the ground and produces fine-grained material.

It is difficult to make an estimation of the rate of the glacial erosion processes. In the majority of glacial areas, traces of the pre-glacial surfaces are easily recognised and, thus, post-Pliocene erosion can be easily recognised. However, in continental glacial districts, traces of the pre-glacial surfaces appear to have been subjected to little alteration, while at the same time there is a complete absence of the soil.

Small forms of erosion

The microforms which are created by glacial erosion are related to the abrasion process, and usually exist in the form of small stripes and friction

Ice marginal lake at Glacier National Park (Canada) (by C. Centeri).

lines. These are generated by the angular fragments transported by the glacier base. In particular, the friction lines are usually parallel to the direction of glacier movement.

Large forms of erosion

The specific landforms derived by glacier activity depend on a variety of factors such as:

- Glacier type, glacier thickness, the speed of glacier motion and temperature of glacier base.
- The bedrock structure, lithology and tectonic status (diaclases).
- The topography.
- Time.

Large glaciers cause rock compression but little or no erosion, therefore, a succession of little hills and ditches can be observed.

There are regions where the glacier occupies more than one basin whilst the intermediate spaces remain intact. This is the case of selective linear erosion which happens in areas of North America and Europe.

Another glacier type usually restricted in the valleys provides typical forms of erosion known as the Alpine glacial relief. In this case, the passage of the glacier causes the broadening and deep erosion of the valley (valley geometry exhibits a U-shape).

Glacial deposits

The loose material that is transported and deposited by glaciers and associated streams of water is called drift. This material is the result of glacial abrasion. Drift deposits are divided in two categories:

- Unbedded drift deposits, which are directly deposited by the ice.
- Bedded deposits, which are produced by watermelt action.

Unbedded deposits

Although these formations are defined as unbedded, usually, there are some distinguishable abrasion levels caused by the continuous advance of the glacier front. Those deposits that are composed of horizons of stratified sand can be defined as tillites and are mainly a mixture of sand, silt and clay (5-50%) and coarse material (usually less than 10%). Some characteristics of tillites are: the great variety in the sizes of rocky components, the absence of sorted material, the striations on the rock fragments, the orientation of the elongated stones, the great compression of the component material and the sub-angular shape of the associated stones.

Furthermore, in several cases, tillites contain material of much larger size which differs in composition from the material found in the bottom of their mass. This material is known as erratics or erratic blocks; they may be deposited in the form of independent blocks on protrusions of uncovered ground protrusions.

The highest percentage (approximately 90%) of tillites component material originates from areas located up to 10 km away from the deposition site. However, there are several exceptions where tillite component material has travelled longer distances (100 – 1,000 km).

The deposits transported directly by the glacier are often characterised by distinguishable landforms which are referred to as moraines. Moraines are mainly developed across the glacier mass and consist of angular stones, gravel and clay. Depending on the location of their deposition site they may be categorised in final, lobe-shaped and retreat moraines. The last moraine type is formed in internal glacier areas which are characterised by the disruption of glacier continuation.

In districts adjacent to fully developed moraine systems, tens or hundreds of elliptical-shaped hills are extended in an area with a total length and height ranging from 100 to 5,000 meters and 5 to 200 meters respectively. These hills are arranged with longitudinal axes parallel to the direction of the glacier movement and are called drumlins. Their formation is due to the erosion caused by the glacier movement on previously deposited material.

The term "drumlins" is used for glacial deposits characterised by a composition similar to the tillites having the shape of a whale back. The length of a single drumlin may reach 1,000 meters and its profile may be characterised by higher slopes as the altitude increases. Sometimes, drumlins demonstrate stratification and their principal axis is parallel to the direction of the glacier movement. They usually appear in groups and the created relief is called basket of eggs since these formations look like a half-egg shape.

Bedded glacial deposits

The highest drainage rates of the water derived from glacier melting occur during summer near glacier margins. These water quantities may create streams carrying sedimentary material of various sizes. As with other fluvial processes, these sediments are gradually and successively deposited, in layers of different forms, as outwash material. The highest percentage of this material is transported and deposited beyond the glacier margin and can be characterised as proglacial deposit. When this material accumulates in a valley or a plain it may be called a fluvial-glacial deposit, and if it accumulates in a marine or lake environment it can be called marine-glacial or lake-glacial deposit respectively.

The passage of a glacier through an area may cause the creation of several lakes which are the result of ice mass melting inside subglacial cavities. A typical example is represented by esker type landforms that are formed by the deposition of material transported by water-streams under the ice cover. Esker formations are wavelike or rectilinear longitudinal ridges which consist of stratified deposits comprising mainly sand and round-shaped stones. They are produced within water-flow beds located under the glacier, from the melting of ice mass when it is immobilized.

Other characteristic landforms in this category are the kames. The kames are deposits characterised by conical shape and are the result of glacier melting which takes place in old river deltas or glacial valleys. They usually are derived by the overflow of lakes situated in front of the glacier mass. They consist of well sorted sands and round –shaped stones.

Kettles which are often found in glacial environments are formed inside small ground depressions

Ancient moraine lake after the glacier's retreat. Moraine lake (Canada) (by C. Centeri).

(they are also characterised as kettle-holes). These depressions usually are filled with water and form the kettle lakes. These terms are mainly used in geological terminology for the subsidence formations created in moraine areas and the abrasion plains of glaciers. The existence of kettles is due to the coverage of frozen land sections by glaciers deposits. When frozen sections melt, the overlaid deposits start sinking.

Areas covered by heavier sediment loads are called outwash plains. The bottom deposits are tightly connected with the surface sediments. Some deposits appear in the form of outwash fans and part of the silt fraction is deposited by the outwash channels, creating silt barriers. Powerful winds in combination with a dry or low humidity environment, may drift tonnes of these deposits and accumulate them as a loess-type of sediment (a loess blanket). The thickness of this typical aeolian deposition ranges from 10 centimeters to 20 meters or even more and covers areas of great extent in the outer glacial and periglacial regions of North America, Europe, and Asia. In some mountainous valleys (mainly in Central Europe), series of terraces, in various levels, can be observed and each of them corresponds to a glacial period. These deposits are very useful for the dating of Pleistocene glacial incidents. Most of the silt fraction is easily transported by meltwater and eventually reaches a lake or a marine environment. In deep fresh water environments, the fine-grained material creates deposits which are called varves. The bottom layers of the varves are light-coloured and

117

represent flood incidents or spring storms. In contrast, the uppermost layers are dark coloured and represent deposition under tranquil conditions during winter time. These varve couplets may have a variable thickness ranging from 1 to 100 mm. Shallow glacial lakes may become covered by salt deposits causing bottom siltation and, hence, interrupting the sequence of the annual varve couplets.

Finally, there are also depositional formations comprised of gravel and sand layers of relatively good stratification and exist near fluvial streams.

Periglacial areas

The areas which are not covered by ice and located near the glacier margins are called periglacial. There, the land topography is greatly affected by low temperatures and the neighbouring ice masses, resulting to the formation of typical landforms. The evolution of these landforms depends on the intensity of glacial influence. In areas characterised by long periods of very low temperatures and short summer periods, there are ground sections which are permanently frozen. This is called permafrost and can reach to a great depth. In high altitudes, when underground water gets close to the surface, within the permafrost zone, there is a tendency for ice formation. In areas, where underground water creates springs, it freezes and forms hydrolaccoliths. Near the surface, this hydraulic forces causes the ground to form a bulge, like a miniature volcano of a height which can reach up to 100 meters. This structure is widely known as pingo in Siberia and Canada; when it melts due to climatic changes, the entire place is covered by a lake.

The daily processes of freezing and melting may lead to the gradual decomposition of rocks, hence, every porous rock becomes particularly fragile. This may cause ground displacement and contribute to the creation of many landforms of restricted size known as patterned ground.

The areas which are located within the glaciation zone but have never been covered by ice are characterised by extreme gelifluxion effects. In some of these areas, during the peak period of glaciation, the development of specific flora and fauna is favoured; they are called glacial ecosystem refugees.

While the ice mass advances, the glaciers tend to interrupt existing branches of the drainage system and form lakes; these may overflow to glacial canals (glacial spillways) having destructive effects. During the glacier retreat, large masses of melt-water form periglacial and proglacial lakes. These lakes are emptied through larger glacial canals and spillways during deglaciation when the earth's crust moves isostatically.

Rock glaciers are blocks of angular coarse-grained material. They look like small glaciers but ice is not their principal component. They are periglacial forms which occur by the creeping of the permanent glacial cover.

Glacial and Eustatic processes

Ice overloading on a continental region always causes compression and sinking of the earth's crust to a depth approaching one third of the thickness of the overlaying ice mass. This external change of the crust shape which may also be caused by other factors , is known as warping.

Likewise, when deglaciation is in progress metaglacial isostatic movements of crust restoration take place. In an ideal system, crust restoration could be completely achieved but, in reality, it is not clear if full restoration can take place.

In coastal areas the original coastlines can be mapped. The use of ^{14}C for the dating of the varve deposits and examination of the organisms found in the deposits of the elevated coastlines, may provide useful chronological indexes. The comparison of these indexes with modern curves defining land altitudes may determine the isostatic curves of equal emergence or submergence.

Beyond the ice sheet margin a discharge of the isostatic tensions is developed. This is the elastic reaction of the earth crust to the initial vertical pressures which were caused by the ice sheet. During deglaciation, it seems that this marginal discharge of the isostatic tensions probably decreases and retreats like a wave, with the regressing ice cover. At the same time, glacial valleys are flooded due to the sea rise level, creating fjords and deep gulfs.

In the primary stages crust restoration takes place at high rates and may last only for several hundreds of years, whilst during the next stages it is very slow and can last for thousands of years. The identified difference in the rate of isostatic restoration may reflect different levels of reduced crust

Glacial lake within an old glacial cirque. When the glacier melts away, a cirque bottom may remain filled with water, making a small, rounded lake called a Tarn. North Cascades National Park (Canada) (by C. Centeri).

resistance.

During glacial periods the volume of the ocean water decreases since it is taken up by the forming ice mass. This decrease results in a global decline of the sea level which is known as the effect of glacial eustasy. Generally, a conversion of $360*10^9$ cubic meters of water into ice corresponds to a global sea level change of approximately 1 mm.

The overloading of the crust brought about by the surplus of sea water in the continental platforms and by the weight of ice masses of Greenland, North America and Europe, has led to great global geomorphological changes. The continuous subsidence of some ocean basins, particularly in the western Pacific Ocean and Mediterranean Sea, has been accelerated during Quaternary period resulting to a lowering of the sea level which is alleged to be approximately 100 m.

The sections of the earth affected by the presence of various ice formations (glaciers, ice sheets, ground ice, sea ice) constitute the Cryosphere. During the Quaternary period, more than 40% of the earth's surface and oceanic areas has been included in the Cryosphere.

Expansion of glaciers during the Quaternary

During Quaternary geological period, several environmental changes have happened but the most severe is the one that ground has suffered by the development of the huge ice sheets. Their repeated progradation and retreat has dramatically affected areas of the Northern and Southern hemisphere.

Furthermore, a secondary impact caused by the existence of the permanent or seasonal ice layer has been the freezing (often reaching great depths) of the soil and underlying rocks or sea bottom material.

The great expansion of Cryosphere is one of the major characteristics of the glacial periods of the Quaternary period. During these periods, large ice sheets have been formed and destroyed.

The advance and retreat of an ice sheet follows each climatic change, but with some delay. This depends on the ice sheet volume, the occupied area (which may be restricted by horsts or mountainous uplifts along its margins) and the nature of the climatic changes. The total area affected during a glacial period can be indicative of the size of the paleo-Cryosphere and the ice volume existant in glacial areas. The total area covered by ice during a typical glacial period at its maximum phase is estimated to be approximately $40*10^6$ km^2 (for comparison, the frozen area before the glacial peak can have an extent of $15*10^6$ km^2) whilst the volume of water which is stored as ice during the glacial peak is estimated to be about $90*10^6$ km^3 (for comparison, the current water volume is about $30*10^6$ km^3). Therefore, it seems that during glacial peaks ice volume can be tripled whilst frozen areas may be extended to regions which are 2.5 times larger than those occurring before a glacial peak. Furthermore, prevailing periglacial conditions may have a significant influence on a given area, either during glacial or interglacial periods. There is still an uncertainty concerning the modelling of the conditions of the

U shaped valley previously occupied by glaciers. Glacier National Park (Canada) (by C. Centeri).

last glacial peak because the ground data have not been completely verified and it is possible that some parts of particular ice sheets are part of a wider system. For the solution of this problem, an understanding of ice sheets dynamic behaviour is required in order to explain the expansion of these sheets in areas near the Equator. Some uncertainty also lies on the issue of ice expanse on the continental shelves located presently beneath sea level. These areas may only be explored with considerable difficulty so their sediments may be mapped and dated inaccurately.

Reasons for the Development and Retreat of the Glaciers

The study of the Cryosphere expanse during the last glacial period and the dating of the various stages of ice advance and retreat, may lead to the explanation of the reasons which caused the glacial and interglacial events.

The precise processes associated with the growth of the most important ice sheets remain undetermined. Milankovitch, in his astronomical theory, argues that variations in the solar exposure of higher northern geographic latitudes during summer seem to have significant contribution to climatic changes. The lowest solar heat supply, defined by the features of Earth's orbit (mainly ellipticity and axis inclination), periodically allows the preservation of summer snow whilst additional reflectivity caused by the existing snow cover (albedo) makes the atmosphere cooler. Therefore, slow snow accumulation could bring about, eventually, the advance of glaciers in mountainous regions of higher

121

Hanging valleys with waterfalls join towards a U shaped valley. Glacier National Park (Canada) (by C. Centeri).

northern geographical latitudes, combined with a gradual expansion of the ice-covered area. Additionally, it is possible that ice advance could have been accelerated when permanent snow margins began to move towards the south, following the temperature decline. This theory was named as the direct glaciation theory. Furthermore, according to this theory, the first places at which ice accumulation started are those in Baffin Island, Labrador, Rocky Mountains, Alps and Scandinavian mountains.

Furthermore, two additional factors have also contributed considerably to the development of glacial periods:

- The existence of high humidity values, which implies the presence of a quite warm ocean in wind's direction.

- The minimum loss of accumulated snow and ice. For example, an internal mountainous area not connected through glaciers with the sea (thus, avoiding the creation of icebergs and the subsequent reduction of snow mass) could be ideal for the development of ice sheets.

Measurements of oxygen isotope concentrations in the sea water which are considered to reflect the global ice volume, have demonstrated that short periods of glacier advance and expansion should also occur in oceans. This finding may be verified by the rapid increase of the ^{18}O values in the fossil foraminiferae dated from the periods of 11,500, 7,500 and 2,500 years B.P. The first and second time periods are the most important, since according to the estimations of Ruddiman et al. (1980), 50% of the total ice volume derived during last glacial period was formed during these periods.

Temperature values estimated from the existing foraminiferae populations, indicate that the first period of ice advance (115,000 years ago) took place before the commencement of the significant cooling of the Atlantic Ocean surface, particularly, in geographic latitudes from 40º to 45º . Therefore, it can be deduced that ice development preceded the oceanic temperature drop. This may be explained by the fact that ice had been developed in areas not connected with the sea, and thus, despite ice accumulation, there was not sufficient ice contact with the sea water to bring about a reduction in the average oceanic temperature. This seems to be confirmed by the theory of inland ice accumulation and agrees with some prerequisites have been mentioned above. According to astronomical measurements, the periods of 11,500 and 7,000 years B.P. were characterised by the lowest solar exposure during summer, particularly, for areas located in the geographic latitude of 70º N. This finding supports the presumption that ice development took place in the way described by the Milankovitch theory.

Modern glaciers

At the present time, 10% of the earth's surface is covered by glaciers and it is estimated that they extend over an area of $14.^9*10^9$ km². The largest glaciers, in terms of covered area, are found in the Antarctic ($12.5*10^9$ km²) and Greenland (1.7×10^9 km²) The glaciers may be categorised as inland and local. The first group includes the glaciers of Antarctic and of Greenland which represent almost the 99.3 % of the total glacier existence (in volume) in earth and the second one all the others. It should be emphasised that if the glaciers of the Antarctic melted, the global sea level would rise approximately at 59 meters above the present one; for Greenland glaciers the sea level rise could be approximately 6 meters.

There is a general belief that inland glaciers were formed when, under appropriate climatic conditions, snow fall occurred reaching the height of permanent snow line and then accumulating in layers of significant thickness. Therefore, there was a process of positive feedback for the creation of glaciers. However, in the long term, the slow downward movement of the glaciers due to the decrease of their volume caused a negative feedback.

main glacial landforms

ARÊTE

Arêtes are sharp edged narrow crests which occupy higher elevation areas within the glacial environment. They usually separate two parallel glacial valleys and their composition is similar to the bedrock. However, they must not be confused with the medial moraines, which consist of transferred material. Arêtes can also be formed during the development phase of two neighbouring cirques when the local bedrock is eroded until only a narrow ridge is left between them.

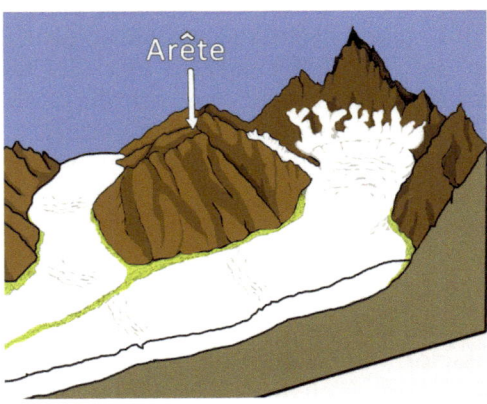

CIRQUE

A bowl shaped landform, which is actually the starting point of a glacier. In glacial environments the cirque belongs to the more elevated formations, along with the arêtes and horns. The three sides of this depression have escarped walls and the fourth side is open and descends into the glacial valley, forming the starting point of the glacier. Before its depression, a cirque appears as a simple irregularity on the side of the mountain, later augmented in size as it becomes more and more occupied

by ice. When the glacier starts to heave towards lower altitudes, the open side of the cirque is widened. After the glacier melting, these depressions are usually occupied by small mountain lakes, called tarns.

Alps (by. K. Pavlopoulos)

CREVASSES

They appear on the surface of a glacier. Their genesis is a result of mechanical processes due to the succession of freezing and melting. Additionally, during the intrusion of a glacial tongue into the sea, the section of the submerging glacial mass is lifted (due to its lower specific gravity) and the fissures are gradually widened, resulting to the detachment of icebergs from the ice body

Clavell glacier-Canada (by C. Centeri)

CRYOTURBATION

Disturbance of the ground caused by successive alternations of melting and freezing.

DIFFUSE FLOW

Meltwater flow occurring in thin layers or amorphous small streams.

Glacier National Park-Canada (by C. Centeri)

DRAINAGE CHANNEL BENEATH THE GLACIER

A pipe or channel in the sub-glacial area which acts as a drainage passage for meltwater. This drainage channel extends up to the glacier gate which is the water's outflow point.

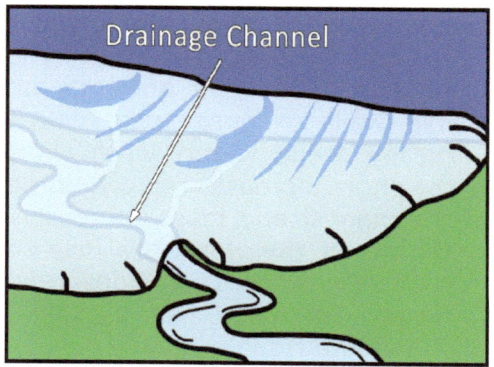

DRUMLIN

Hill of moraine deposits of elliptic shape, characterised by similar with the moraine material arrangement. A drumlin has the shape of a whale's back. It is located under the glacier and may have a rocky core. Drumlin dimensions may vary from tens to hundreds of meters, with their width being smaller than their length, and height ranging from 5 to 40 meters.

Glacier National Park-Canada (by C. Centeri)

FIELD OF DRUMLINS

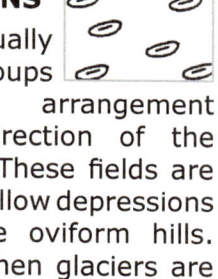

Drumlins usually appear in groups with longitudinal arrangement parallel to the direction of the glacier movement. These fields are characterised by shallow depressions which separate the oviform hills. They are formed when glaciers are very rich in moraines and silt due to the relatively high erodibility of the glacier valley.

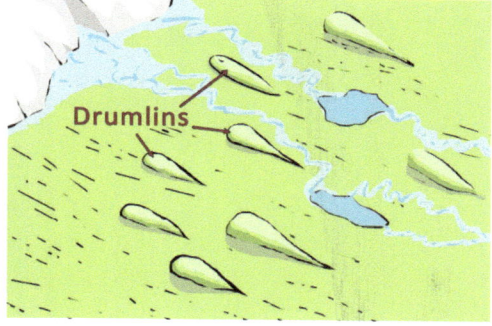

ERRATIC

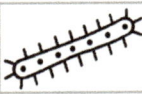

Rock block, located hundreds of kilometres away from the nearest appearance of the respective bedrock (allochthonous origin). The theory that erratics have been transported by ice has overcome the older theory which argues that these large sections have moved during biblical floods or detached by big floating icebergs. The residual parts of the erratics are located abandoned near the margins of a regressing glacier.

Alps (by. K. Pavlopoulos)

FIELD OF ERRATICS

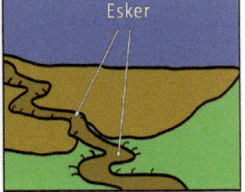

Area characterised by the presence of dispersed boulders transported by gelifluxion or ice.

Alps (by. K. Pavlopoulos)

ESKER

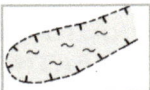

The term is Irish and corresponds to an alluvial formation which consists of sand and gravel material in alternation. It is a narrow and long structure located inside a glacier's tunnel or under the glacier and becomes apparent after glacier's regression. Its direction is indicative of the ice motion. These forms are created by meltwater activity underneath the ice sheet and their height varies from one meter up to tens of meters with their length ranging from hundreds of meters up to kilometres. Eskers are often used as reservoirs of barren material designated for construction.

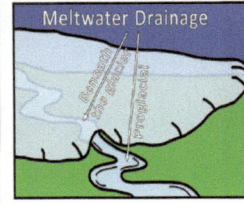

FJORD

Fjords are sea gulfs which are created because of marine transgression which results to the flooding of glacial coastal valleys of characteristic U form. The length of a fjord can be more than 200 m. and its depth more than 1000 m. The height of their steep coasts can reach up to 1000m. The flat floor of the transgressed glacial valley is located far underwater, and thus the visible walls of fjords rise almost vertically, while water depth close to the shore increases rapidly. Some of the biggest and most impressive waterfalls of the world are located in such valleys. Areas of widespread fjords are Greenland, Norway, Chile, Scotland and New Zealand.

Norway (by A. Danilidis)

Glacier National Park-Canada
(by C. Centeri)

GELIFLUXION

Gelifluxion is a type of ground flow and is generated by deglaciation. It is the slow movement, on a slope, of a surface water saturated ground layer which flows on the frozen subsoil.

GELIVATION

Rock breaking caused by ice. The result of frost activity in the meeting points of neighbouring rocks.

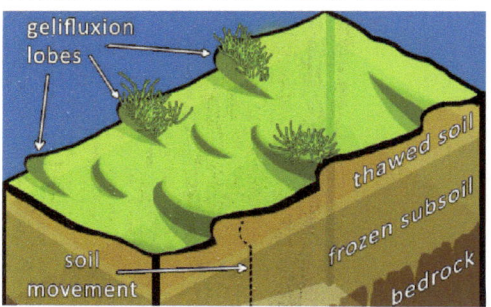

Clavell glacier-Canada
(by C. Centeri)

GELIFLUXION IN LOBES

When gelifluxion is intense, the creeping soil creates lobe-shaped protrusions on the slopes. The extra soil which is accumulated on these lobes is ideal for the development of vegetation.

GLACIAL DEBRIS

Debris accumulated on a slope during the gravitational fall of glacial roundstones.

Clavell glacier-Canada
(by C. Centeri)

GELIFRACTION (CRYOCLASTITES)

Fragmentation of the cohesive rocks due to the successive alternations of glaciation and deglaciation, leads to the formation of debris which are named cryoclastites after their genetic process.

GLACIAL GORGE

A gorge is formed due to the erosion of rocks

127

by glacial meltwater.

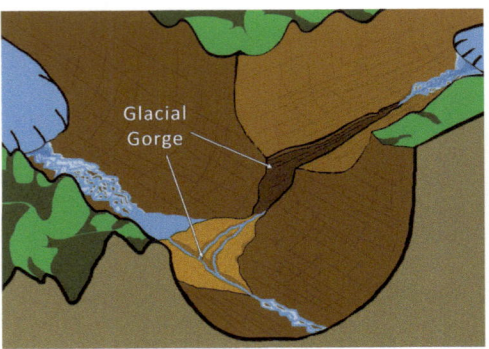

GLACIAL LAKE

It is a lake which occupies a notch or is created when glacier's advance is stopped by an obstacle. During glacier progradation through a fluvial valley or cavity, lakes may form gullies resulting in their expanding while they may be blocked downriver. These lakes usually have an elongated shape and are found in places previously covered by huge ice masses (e.g. mountains).

Clavell glacier-Canada (by C. Centeri)

GLACIAL STRIATIONS

These may be caused by friction forces which glacial valley walls exert on the rocks during their transportation by the glacier. The glacial striations are also apparent

on the valley walls. The rugged and hard relief of the valley ground is rendered smooth with striations parallel to glacier movement.

Athabasca glacier-Canada (by C. Centeri)

GLACIER

Block of ice which is compacted in the form of successive layers. Surface ice melts intrude into the ice body where they re-congeal. During ice compaction in layers, the pressure of the upper layers forces the air which is trapped inside the ice body to escape. This block of ice is not rigid and moves towards lower altitude levels due to gravity forces. The movement of glacier takes place through the processes of fragmentation and re-coagulation which provide the glacier with the feature of plastic mass.

Athabasca glacier-Canada (by C. Centeri)

GLACIER BORDER

It is the ice sheet frontier and beyond that no ice covers the ground. The limit of ice progradation may be defined by the debris forming the glacial tillites and the moraines (tillites mounds) which very often act as barriers. Important information for the study of ice sheet development can be derived by the deposits of the glacial leaching i.e. the material which is transported by meltwater streams developed near glacier margins. Usually, that material is deposited in lower areas of the valleys or in other sections of the glacier margins.

the, glacier advances. When ice melting is faster than ice integration the glacier retreats and is called recessive. When ice integration and melting are taking place at equal rates the glacier tongue is in equilibrium. In polar areas, inland glaciers move towards the sea. When an ice sheet enters the sea, it has the tendency to float due to its lower specific gravity compared to that of sea water, resulting in its fracturing. This is the mechanism of iceberg formation.

Athabasca glacier-Canada (by C. Centeri)

North Cascades National Park-Canada (by C. Centeri)

GLACIER GATE

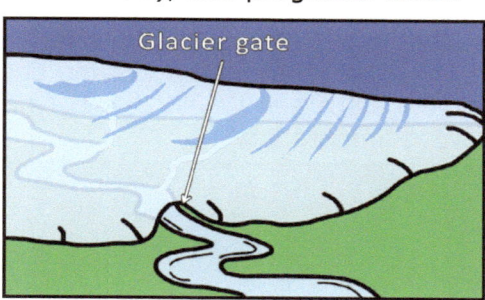

The spot of meltwater spring on the glacier front. After its exit meltwater creates proglacial channels (usually of diffuse flow), and proglacial lakes.

HORN

The meeting point of three or more arêtes. It has the shape of a pyramidal peak with extremely steep sides. This peak is usually the highest point in the local glacial environment.

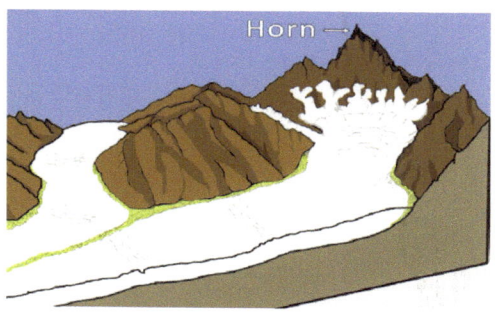

GLACIER TONGUE

It is the front part of the glacier, also known as glacier front. When ice integration is faster than ice melting, the tongue moves downslope; therefore

KAME

Deposit on the margin of a glacier within a

notch formed by the glacier. Usually it occurs in the form of hills of poorly sorted sand and gravel. In rare cases, it consists of tillite and silt which has been deposited by the flowing meltwater.

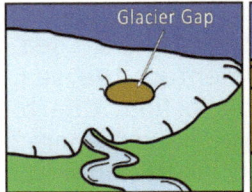
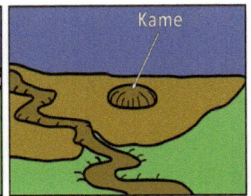

KETTLE HOLES

They are little depressions on the surface of glacial grounds. When they are filled with water they are referred to as kettle lakes. The term kettle holes is also used for depressions within moraines or the outwash plains associated with them. For the first case, the depressions may be 30 to 300 meters wide and 10 meters deep, while for the second case, their width may reach up to some kilometres and their depth up to 30 meters. Their generation in both cases is caused by the covering of frozen ground sections and ice blocks by glacier deposits. Melting and recession of these sections enhances the depression of the overlaying deposits, thus, generating these holes.

LOESS

They are aeolian depositions of silt in areas which are characterized by a cold periglacial climate. It is a characteristic sedimentary formation at periglacial climates and consists of various unstratified or very fine-layered components alternated with sand and gravel which are produced by ground flow processes. Various forms and layers of loess have been studied and in whole they make the loess sequence.

Galgaheviz-Hungary (by C. Centeri)

MILL OR MULLIN

This is a drainage pipe within the glacier which begins from the glacier surface and discharges meltwater through the glacier's body. The thaw of the upper sections of the glacier produces meltwater which travels through mill down to the glacier's foot, where it comes out through the glacier gate as a torrent. Initially, meltwater is characterised by turbidity due to its high content in silt and sand but later it becomes limpid due to the low sedimentation.

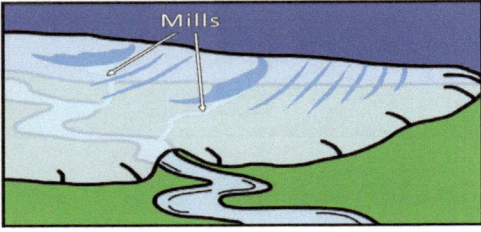

MORAINES

They are glacial deposits, generally, consisting

of heterogeneous coarse-grained material. The almost unsorted debris which composes them very often acts as a barrier which defines the limit of ice progradation. Moraines may be classified acording to their position. A sub-glacial moraine, characterised by vast heterogeneity, is called a ground moraine. It is argillaceous with pieces of variable size, characterized by exceptionally intense friction, and it is a very fine-grained glacial aleurite. A moraine located within the glacial valley, on the sides of a glacier, created by the debris coming from the valley slopes, may be defined as lateral moraine. A moraine formed by the conjunction of several moraines located between two parallel glacier masses, is called a medial moraine. In addition, the moraine which is formed before the glacier front is called a front moraine. Finally, when the glacier starts to retreat, the provision of material to the front moraine is terminated. In this case the front moraine is converted to the terminal moraine of the glacier.

Hills of moraines is the relief created by moraines material which is accumulated in the form of hills. The formation of moraine hills is related to various processes, such as the melting of glaciers, tectonic activity or glacial-isostatic movements.

Lateral Moraine

Lateral Moraine

Ground Moraine

NEVE, FIRNSCHNEE

This is considered as the primary stage of glacier formation. The upper section of the accumulated snow mass melts and the produced water intrudes into the unfrozen part of the ice body and refreezes. This process results to the formation of neves which are a mass of accumulated crystal granules. While granular ice is developed, an effervescent structure is built due to the air trapped in the snow crystal's needles. Then compaction proceeds resulting in the formation of the glacier.

Tatra mountains-Slovakia (by C. Centeri)

POLYGON OF STONES (PINKO)

A polygon-shaped structure made from stones usually presented in groups with small distances among them. They are created by fragments produced by friction due to the pressure of ice load and the effects of melting and freezing. Each polygon is characterised by an accumulation of rocks in its circumference and an intermediate space which consists of smaller size components.

PROGLACIAL CHANNELS

These are drainage channels formed before the glacier tongue that carry the flow of meltwater after its exit from the glacier gate.

Banff National Park-Canada
(by C. Centeri)

PROGLACIAL DEBRIS CONE

A debris cone formed by a water stream which can be fluvial or proglaciall.

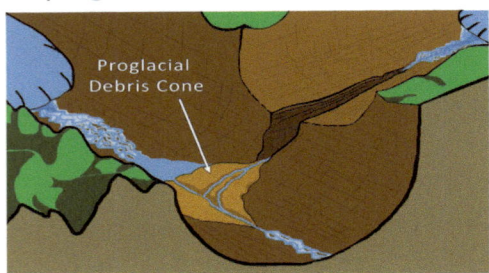

Proglacial
Debris Cone

ROCK GLACIER

Moraine material mixed with ice, characterised by crevasses and melting ice cores. The presence of rock glaciers indicates the retreat of a glacier and it is an evidence of its final extinction.

Athabasca glacier-Canada
(by C. Centeri)

SERACS

Ice pieces of chaotic structure, unstably located on the surface of a glacier. They are created during crevasses formation, due to the acceleration of ice sheets while they move on slopes of high inclination.

National Park Los Glaciaires-Argentina
(by S. Kanitscheider)

Clavell glacier-Canada (by C. Centeri).

Samos Island - Greece (by A. Vassilopoulos, N. Evelpidou)

Chapter 6

karstic environments

karstic processes

Karst-Introduction

The term karst derives from the Slavic word Křs, which is the name of a limestone region in Slovenia and signifies a rock formation or a rocky area.

According to other researchers, the term derives from the German word Karr or the Italian Carso, which characterises an area of white limestone, poor in vegetation, located north-east of Trieste.

The German name of the region Karst has been established by the international geological and geomorphological bibliography and has been defined as a scientific term that refers to areas that present relief similar to the one of Kras region in Slovenia.

Karstic areas have specific relief and drainage features, developed on rock formations highly soluble to water. However, it represents more than static features. Karst is a dynamic system of landforms, life forms, energy, water, gaseous and solid substances, and bedrock. The perturbation of any of the system's factors affects the rest of them.

The karstic relief is characterised by a variety of forms, which depend on the climatic conditions and the rocks' nature. The karstic area usually appears dry, lacking surface runoff and vegetation, despite the prevailing high rainfall, due to underground water drainage.

The subterranean runoff is usually rapid and there is a deficiency in the filtration of natural water pollutants and suspended particles, and thus pathogenic organisms survive in underground waters.

Unlike karsts of tropical areas, where a surface hydrographic

Karren landforms in limestone formations (Erymanthos, Greece) (by N. Tsoukalas).

network is developed, seasonal surface runoff rarely exists in dry and semi-dry areas.

Aside from limestone and dolomitic rocks, karst forms can also be developed, mainly in dry climates, on the extremely soluble evaporites (gypsum, anhydrite and halite). Karstic forms, in extremely humid climates, are formed on less soluble rocks, like quartz-diorite in North Colombia and eclogite in South Guinea.

The fundamental condition for karst formation is the rocks solubility in water. Solution acts in various ways, but its most important action lies in the continuous amplification of cavities within the rock, its increasing permeability and its continuously increasing ability to pass all the way through it, great quantities of water. The above mentioned actions result in the development of an underground drainage network that causes separation of valley systems and vast areas of karstic caves.

The rocks' behaviour in various karstic landform processes depends to a significant extent on the permeability, mechanic tolerance and purity of the rock, which in their turn mainly depend on the rock's porosity and discontinuities.

Forms that resemble the karstic ones but are produced through different processes and not by dissolution are called pseudokarst. They usually appear on sediments of fine-grained pyroclastic material, which condense because of the water (e.g. the weathering of silicate-argillaceous minerals generates landslides and formation of cavities due to mechanics), on bedded lavas (water infiltration on the layers'

Gours landforms at Mainalo (Greece) (by N. Tsoukalas).

surfaces), on permafrost, due to local melting, as well as on granitic rocks under the form of large circular cavities called tafoni.

The rocks that are mainly subjected to karstification are: limestones, dolomites and evaporites, which represent 30% of the earth's surface.

Dissolution of limestones

Rainwater cannot dissolve limestones' $CaCO_3$ by itself. During its fall, however, gets enriched with atmospheric CO_2 according to the following reaction:

$CO_2 + H_2O \rightleftharpoons H_2CO_3$ (carbonic acid)

therefore, the resulting carboniferous water takes effect on limestone rocks and produces calcium bicarbonate:

Karstification processes in carbonate rocks led to the formation of Vouliagmeni lake. South Attica (Greece) (by A. Vassilopoulos, N. Evelpidou).

$$CaCO_3 + H_2CO_3 \rightleftharpoons Ca(HCO_3)_2$$

The calcium bicarbonate that is produced by the previous reaction is 30 times more soluble, in pure water, than calcium carbonate. The dissolution of limestones is facilitated a great deal in tectonically stressed areas. Cracks and fissures, as well as faults, facilitate the passage of water through the rock as well as its "in depth" dissolution. This process, which is similar in other types of rocks, has continued for thousands of years and is responsible for the formation of the surface and subterranean karst.

Almost all dissolution processes are altered by factors that act on the surface and at a small depth. The surface vegetation adjusts karst's water flow through the root system that withholds water, water's infiltration in the soil and the production of CO_2 and carbonic acid. Water retention from trees and their root systems affects the quality of water that will contribute to the karstic processes. Trees release in fact 20-25% CO_2 through their root system.

Karstic geomorphology

Karstic geomorphology is the branch of the science of geomorphology that deals with the study of karstic forms. Within the framework of karstic geomorphology we may identify, map and study surface and subterranean karstic landforms.

Karstification is described as the process of chemical weathering (dissolution) of carbonate rocks' (limestone, marble etc) and evaporites' by water . This results in the formation of typical surface or subterranean landforms called karstic landforms.

Karstic landforms are morphological formations resulting from the dissolving action of water on particulary soluble rocks. Such rocks are limestones, dolomites and evaporites. Karstic landforms may appear on other rock categories, but only rarely and under specific climate conditions.

According to the theory of Petrović and Živago, the evolution of karst and karstic relief depends on the hydrography, lithography, and climatic features of an area and is divided in the following phases:

• *Phase I*:Concerns the pre-karstic cycle of erosion in humid areas. The evolution of relief, in its primary stages, is gentle and gradual on carbonate rocks. In this phase, the soil's surface is permanently humid, vegetation maintains a natural evolution, cracks are absent or very few but of low density and usually covered by vegetation and soil mantle or colluvial deposits. The underground aquifer is relatively shallow and that is why surface runoff is important.

• *Phase II*: In this phase the cracks start to open. The cracks absorb most of surface water, thus facilitating and accelerating its circulation inside deeper aquifers. In this way the karstic dissolution continues deeper and the level of the aquifer goes down. During phase II riverbeds are deepening and karstification becomes obvious.

• *Phase III*: In this phase the cracks have been expanded, small channels occur (clints and grikes) facilitating significantly

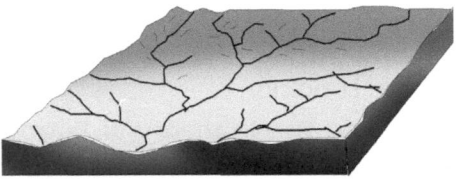

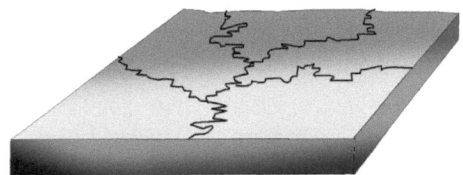

Davis karstic evolution cycle.

the movement of surface waters, so that they flow into the deeper layers even quicker. The hydrographical network (rivers and watergullies) is periodically drained (dry karstic valleys) introducing a seasonal flow. Small karstic formations, such as clints, grikes and dolines begin to appear on the bare karstic relief. Groundwater is moving towards greater depths and karstic springs appear around the dissolvable carbonate rocks.

- *Phase IV*: At this phase cracks have been expanded and deepened considerably. The hydrographical networks lead to sinkhole systems and exhibit low and seasonal surface runoff. Instead of surface fluvial gullies the subterranean karstic network is intensely developed, whereas some parts of hydrographical networks go out of use and are "fossilised".

- *Phase V*: In this phase, along the gully beds, in altitudes lower than those of sinkholes, dolines appear. Former river mouths now emerge as hanging valleys. Blind valleys are formed. The subterranean karstic tubes of the subterranean rivers have already been considerably expanded and big subterranean caves have been formed.

- *Phase VI*: This is the last evolutionary phase, in karstic areas. This phase begins when in the blind valleys sinkholes appear and suppress the hydrographic networks' underground flow; thereby even the subterranean rivers becomes dry and "fossilised". The adjacent dolines of the previous phase are now linked together and have become uvalas. The permanently humid and water saturated (wet) zone is now located deeper and in the non-saturated (dry) zone continuous and dominant karstification takes place. Some caves' roofs fall and so they open, exposing the subterranean networks and karstic tubes to atmospheric processes.

- *Phase VII*: In this phase karstification is interrupted. The surface karstic forms are transformed or are covered by newer sediments, while the subterranean karstic forms, passages and tubes are associated in more complex ones, and are influenced by the processes of the previous pro-karstic cycle.

Forms of dissolution

There are numerous karstic landforms that vary in shape and size. They are divided into surface and underground forms.

The clints and grikes, finger marks and rills, belong to the surface forms that were formed on the surface of limestones because of the activity of the rainwater. When dissolution proceeds in depth, usually supported by the presence of cracks, it creates cavities of great depth with vertical walls, called vertical shafts.

Dolines, uvalas, and poljes are important karstic macroforms.

Dolines are closed basins of relatively small dimensions (5-20m deep and 10-1000m wide) of circular or elliptical shape with larger width than depth. They usually occur in groups, and then compose a "dolines range" and provide the area with a particular morphology. Their appearance on flat surfaces of specific altitudes constitutes evidence of a planation surface. The comparison of doline dimensions at various altitudes (planation surface) provides significant information about the relevant tectonic movements of the area.

Uvalas are karstic landforms that result from the association of two or more dolines, that is to say from their amplification, due to the continuing dissolution. Consequently, uvalas constitute an evolutionary stage of dolines.

Poljes are closed basins of great dimensions that have the shape of a valley. Their floor is almost flat and covered by alluvial deposits, mostly clay material that is the residue of limestone dissolution. The flat surface of the polje floor is frequently interrupted by hills whose height can reach 100m. They are typical residual karstic landforms called hum and are the residue of limestone's dissolution because of the different composition of the rock at that particular position. There are poljes whose drainage occurs on the surface through a fluvial stream flow,

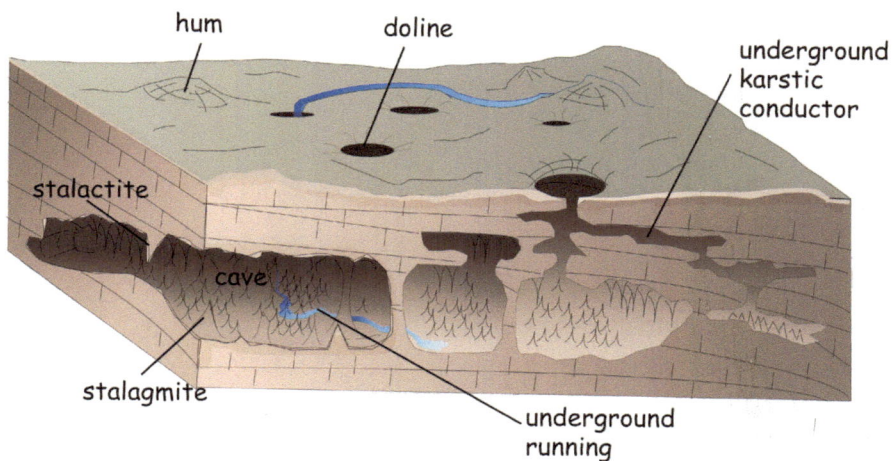

Surface and subterranean karstic landforms that result from the dissolution of limestone rocks by the atmospheric water.

while there are cases of poljes where their drainage occurs underground through sinkholes.

Sinkholes are oppenings on the earth's surface connected to an underground karstic tube system. Once the sinkholes are filled with clayey material they are blocked and as a result water cannot find an outlet and part of it or the whole polje is filled with water, forming a lake.

Underground karstic landforms consist of the undergound karstic tubes and caves, together with a large number of smaller forms (stalactites, stalagmites, columns etc) that are features of their interior.

Karstic evolution cycle

Over time karstic areas "evolve", passing through different stages, known as stages of the karstic cycle.

In the framework of an area's karstic evolution cycle, certain relief features that determine the stage of the karstic cycle that is observed.

Description of stages of the karstic cycle follows the order of construction of surface and underground of karstic landforms.

In the initial stage or the stage of youth, water's solvent activity on the surface of limestone creates clints, grikes, vertical shafts and periodical dolines, while the area is drained by a surface hydrographical network.

During the stage of maturity the surface forms are being expanded and thus uvalas and poljes are formed, while a subterranean drainage network is replacing the surface hydrographical network. The landforms of the subterranean karst,

such as caves and karstic tubes, are also well developed in this stage.

In the stage of senility all karstic landforms have been weathered due to surface erosion. The karst cycle is usually completed with the exposure of the impermeable formation that lied beneath the limestone, due to the latter's complete dissolution. The topography of a relief in this stage, in a karstified area, does not differ significantly from the "theoretical peneplain" of the end of all erosion cycles. The reappearance of surface hydrographical network that was developed underneath the limestones is typical.

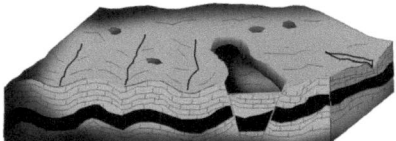

STAGE OF YOUTH

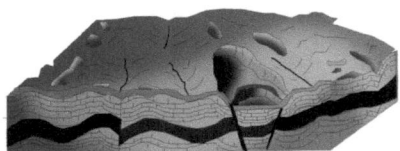

STAGE OF LATE YOUTH

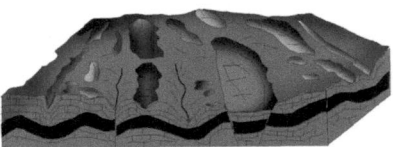

STAGE OF MATURITY

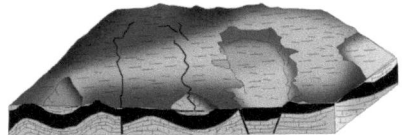

STAGE OF ANILITY

Stages of an area's karstic evolution cycle.

Stalactites and stalagmites inside a karstic cave. Samos (Greece) (by A. Vassilopoulos, N. Evelpidou).

main karstic landforms

CAVE

Caves are cavities of  the ground that have been created in the rocks' interior and which communicate with the Earth's surface through small orifices. Most caves are underground karstic forms. Caves are the largest category of subterranean karstic forms. For thousands of years they have accommodated humans, so that the evolution of the human race depended on them for a long time. Limestones are the most suitable rocks for the creation of caves. The accumulation of the water's dissolvent action in certain locations leads to the creation of small cavities forming caves when they are broadened. However, porous limestones are not capable of forming such landforms, because they allow free intrusion of water in any direction and their solution takes place in a symmetrical way. Usually, under the entrance of caves a pile of roof material is found, the collapse of which resulted to the cave's communication with the surface.

DOLINE

They are the most  common landforms observed in carbonate formations in karst fields. Dolines occur either isolated or in groups. Their generation is due either to the collapse of a subterranean cave's roof, in which case they are called collapse dolines, or to the chemical dissolution of the rock, in which case they are called dissolution dolines. Their creation is favoured by the existence of diaclases, as happens with all karst landforms. Usually small dolines are funnel-shaped with flat bottoms. In that case dolines are considered to be in advanced karstification stage, since depthwise solution that cannot be perpetual, has stopped due to the presence of resistant formations.

Samos-Greece (by A. Vassilopoulos, N. Evelpidou)

Crete-Greece (by K. Pavlopoulos)

BLIND VALLEY

Closed valley located at the dead-end of a stream around a cavity where it disappears.

CLOSED DOLINE

A doline not connected  to the drainage network of neighbouring valleys.

OPEN DOLINE

A doline generally interconnected in a valley network.

ESTAVELLE

A pothole functioning alternatively and temporarily as a sinkhole or a karstic spring.

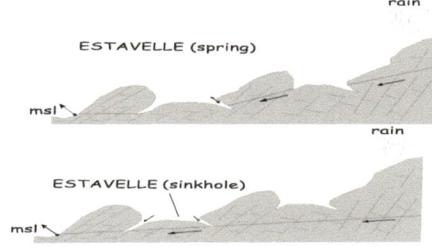

HUM

A residual landform that occurs in karst areas i.e. within poljes. The Hums are calcareous hummocks which represent the residues of karstified limestones.

Evia-Greece (by K. Pavlopoulos)

KARREN, SCULPTURE

They are small karstic forms which occur in soluble rocks. They are divided in free sculptures, semi-free sculptures and covered sculptures depending on the cover of the rock in which they are developed: naked, partially covered or with vegetative or soil cover respectively.

Kopaida-Greece (by A. Vassilopoulos, N. Evelpidou)

KARST

A type of relief, with a specific drainage network, which occurs from the dissolution (karstification) of carbonate rocks.

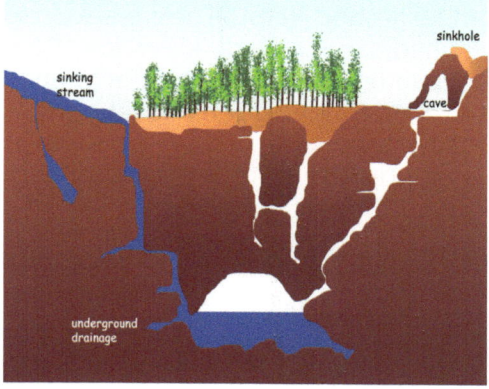

COVERED KARST

A karst surface that lies buried under a cover of laterites and/or under a formation of transported allochthonous material.

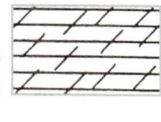

EXHUMATION KARST

Fossil karst that has been uncovered through erosion processes.

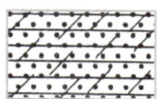

FOSSIL KARST

Old karst that lies buried within a geological formation (i.e. sedimentary), and which can be uncovered by current erosion.

PSEUDO-KARST

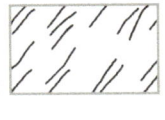

It is a relief which is characterised by landforms similar to karst which, however, are not a result of karstification processes, but of other processes (e.g. chemical erosion in non-karstic fomations).

UNCOVERED KARST

Karst surface constantly exposed to atmospheric processes.

KARSTIC SPRING

They are divided in two main categories, headsprings and springs of underground karst. Their creation is caused either by local elevation of the karstic level, or by the interference of impermeable material (clay, marls) resulting in the increase of pressure. Pressure is rising due to the stuffing of gaps with calcareous deposits from the precipitation of crystal sediments as gypsum, dolomite, calcite etc, occurring during the warm periods (in these periods the concentration of salts in the circulating underground waters increases).

Kopaida-Greece (by A. Vassilopoulos, N. Evelpidou)

of the carbonate rocks, or through the karstic channels, or through the combination of the aforementioned, and outflow below sea level, due to the altitudinal difference. Fresh water concentrations floating on sea water are often created. This effect is due to density differentiations. The lenses of the fresh water on sea water are maintained, if the speed of the fresh water, which supplies these lenses, is higher than the diffusion of the salts of sea water to fresh water. Thus, three zones of different water quality can be distinguished: "fresh floating waters", "Subsaline intermediate waters" and "Sea or Salty waters".

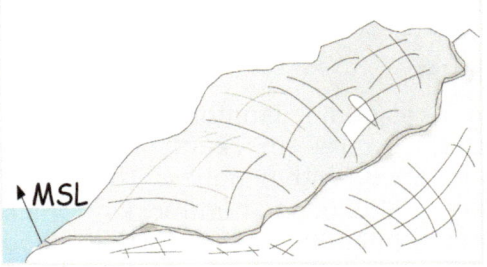

SINKHOLE

It is a karstic semicircular hole connected with the processes of caves' creation.

Kopaida-Greece (by N. Tsoukalas)

SUBMARINE KARSTIC SPRING

The waters of precipitates infiltrate in great depths, either through the diaclases

KUPPEN

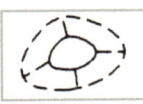

A relief that has a large base and is arched on top.

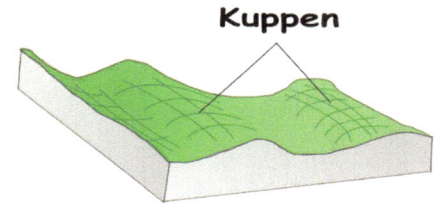

Kuppen

PLATES (STONE PAVEMENT)

Planes or boundaries traced on stones that reveal the uncovered rock.

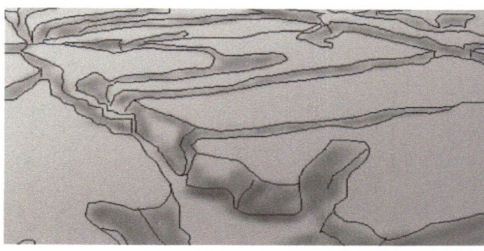

POLJE

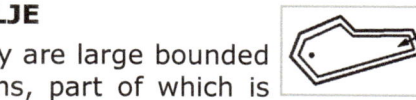

They are large bounded forms, part of which is developed in soluble rocks. They seem like valleys or basins due to their great width and length. The circumferences of these karst plains is steep, their bottom is flat and their drainage is subterranean. Their bottom is covered by fertile soil of "polje type".

Crete-Greece (by K. Pavlopoulos)

OPEN POLJE

A polje generally interconnected in a network of valleys.

POTHOLE

It is an absorbing orifice located within a doline or a polje and is the main drainage path for surface waters. It is created by solution, particularly in areas where faults exist. Potholes lead towards the interior of the rock and form a system of subterranean channels, galleries or caves, usually of labyrinth form.

Kopaida-Greece (by A. Vassilopoulos, N. Evelpidou)

SPRING VAUCLUSIENNE

Reappearance of an underground flow through a siphon, which distributes the water load in a regulatory way.

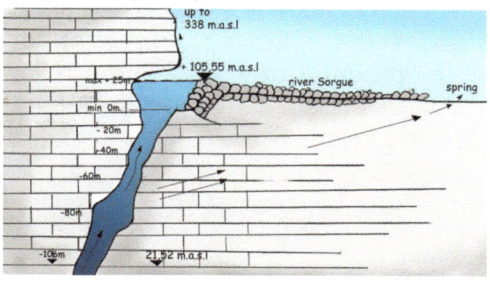

STALACTITE

A typical form of cave decoration due to the accumulation of $CaCO_3$. It maintains the form of a column or a curtain developed from the roof to the floor of the cave. Stalactites are located at points where waters flow in the cave, either through diaclases, or through faults located on the roof. The water flows through the roof in drops, which, while advancing, deposit small quantities of $CaCO_3$. The deposition is very slow, and for that reason the creation of a single stalactite can last centuries or thousand of years. Column stalactites have a small pipe in the centre of their body, which is the path the inflowing water follows.

UVALA

A cavity which has been created by the junction of many dolines.

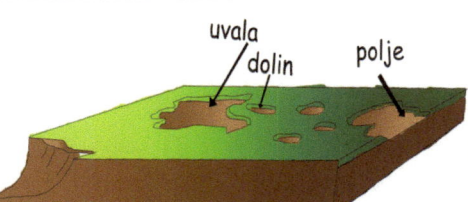

Samos-Greece (by A. Vassilopoulos, N. Evelpidou)

STALAGMITE

A typical cave feature created by the deposition and compaction of $CaCO_3$. Usually created right below stalactites, they are developed in a direction from the cave's floor towards the roof. Besides the opposite direction of development, they also differ from stalactites because they lack the central pipe.

Samos-Greece (by A. Vassilopoulos, N. Evelpidou)

Karstic lanforms inside a cave in Samos Island (Greece) (by A. Vassilopoulos, N. Evelpidou).

Santorini Island - Greece (by A. Vassilopoulos, N. Evelpidou)

Chapter 7

volcanic environments

volcanic processes

Volcanism

The term volcanism refers to the phenomena and the activity that are connected with the ascent and the ejection of the igneous material from the earth's interior to the surface. A volcano as a landform is the point of the earth's surface from which magma is shed after an explosion.

The expression of the volcanic processes depends on the way the magma comes in the surface and its composition. The ferociousness of the explosions results from the chemical composition of the magma. Magma that contains a big proportion of silicon (SiO_2>65%) is called acidic and is characterised by the extended presence of gases and a low density. When these magmas approach the surface, their temperature reaches 900°C and the explosion occurs when their pressure exceeds the weight of the overlying formations.

On the other hand, when the silicon proportion is relatively small (SiO_2<50%), the magma is called basaltic, its explosions are less violent than those of siliceous magma and it has a temperature around 1200 °C at the surface.

The general characteristics of a volcano are:

- The conduit (or pipe), which is the tube that carries the magma from the earth's interior to the surface and usually follows large faults.

- The crater that represents the opening of the conduit to the

Layers of pyroclastic material due to volcanic activity in Santorini Island (Greece) (by A. Vassilopoulos, N. Evelpidou).

surface.

- The volcanic cone that is formed by the explosion or the outflow of the volcanic material.
- The caldera that is formed by the collapse of the crater and usually has elliptic shape.

Even though volcanism is a global phenomenon, volcanoes appear to group together in specific geographical zones. These areas are the boundaries of the lithospheric plates. When two plates diverge, magma of basic composition (with Si content from 46% to 53%) rises from the mantle and is cooled in the oceanic floor. Volcanoes also appear in areas where lithospheric plates converge or in areas where the plates move transversely. In the former case, the rocks of the earth's crust sink to great depth, melt and are transformed into magma that rises to the earth's surface. The lava of these volcanoes has a greater silicate composition, originating straight from the mantle. Volcanoes that lie in the centre of a plate are located over "hot spots" and are characterised by the presence of a stable central magmatic flow from the mantle. The Hawaiian volcanoes are a typical example.

Types of volcanic explosions

Not all volcanoes exhibit the same behaviour. For example, when the lava is basic it flows easily and the volcano's activity is usually calm. The lava gradually fills the volcano's crater, runs over it and eventually may flow to a significant distance, while at the same time gases are being released.

If the lava solidifies the crater will be sealed, causing the interruption of

volcanic activity for a period of time. During this period, the aggregating gases below the crater reach high pressures until the crater's cap blows with an explosion or a series of explosions. The released gases eject liquid and solid materials that are carried to great heights. All these products from the explosive activity of the volcano are called pyroclastic material and are deposited in consecutive layers around and away from the volcano.

There are four different types of volcanoes according to the explosion characteristics, the products of the explosion and the gases pressure:

- *Shield volcanoes or Hawaiian type volcanoes:* The volcanoes of this category deliver fluid lava of basaltic composition for a long time. Their shape is that of a wide cone with a very broad base. Inside the crater there is a lava lake that boils. Usually, the magma's gases are easily released when they reach the surface, thus the lava flow is smooth. The most characteristic examples of shield volcanoes are in Hawaii. Mauna Kea volcano is considered as extinct (the last explosion is dated 2000 years ago) in contrast to Mauna Loa and Kilauea. The Mauna Loa crater has a diameter of 5 km, a slope of $4° - 5°$ and its height is 4194 m above sea level (or 9000 m above the ocean floor). The conical relief continues under the sea surface and ends in a base of 4000 km diameter. Basaltic volcanoes are also found in India, Brazil, South Africa and Antarctica.
- *Strobolian type:* The volcanic activity of this type of volcano is more explosive than the previous

Thermal springs at Santorini Island (Greece) (by A. Vassilopoulos, N. Evelpidou).

one and the basaltic lava doesn't flow as easily as in the Hawaiian type and hardly ever runs over therim of the crater. The abrupt release of gases causes periodical or irregular explosions that are often violent and can destroy the volcano. The main characteristic of strobolian type volcanoes is the formation of interbedded layers of pyroclastic material and lava around the crater. The slope of the volcanic cone is higher in this type than in the previous one and varies around 30° – 45°. These volcanoes can be found in Stroboli island, north of Sicily (Italy), the place where their name came from, Etna (Italy), Erebus (Antarctica) and Fujiyama (Japan).

- *Volcanic type:* The lava is more viscous than in the Strobolian type, while the explosive activity is more violent and regularly destroys part of the volcano. Every

volcanic activity of this type ends with the cone being sealed due to solidification of the magma. In periods of violent explosions, the old crater is completely destroyed and in its place a topographic depression is created that is called an explosion crater. After every explosion activity, large gas and dust clouds are released at great heights. The crater of volcano of the Volcanian type consists almost exclusively of pyroclastic materials. Volcanoes of this type can be found in Italy (Vulkano island, north of Sicily, Vesuvius), Krakataou and Bezymianny (Kamtchatka).

- *Pelean type:* This type of explosion was observed during the activity of Mount Pelée in Martinique, on the 8th May 1902. The main characteristics of this type are the presence of a burning cloud, of temperature up to 800 °C, flowing down the side of the volcano and the formation of an obelisk shaped cap of viscous lava, in the place of the older crater. In the case of Pelée, the cap height was 400 m. It gradually fell apart as the hours went by. The formation of the obelisk shape cap was followed by a discontinuous explosive activity and the release of ash that covered the town of St. Pierre. The existence of the burning cloud isn't followed necessarily by the formation of a brochette, as for example in the St. Vincent volcano, 144 km south of Martinique, where a radiating outflow of superheated gases took place. This type of explosions is also observed in many volcanoes in Philippines.

Apart from the distinguishing of volcanoes according to their explosion characteristics, they can also be classified in the following three categories depending on the time of the volcanic activity.

- *Dormant volcanoes:* The volcanoes that have not shown any activity from the Pleistocene until today.

- *Extinct volcanoes:* The volcanoes that have not exploded in historical time.

- *Active volcanoes:* The volcanoes that still continue to explode.

The consequences of volcanic activity can be classified as primary and secondary.

Primary consequences include all the direct results of lava flows, gases release, mudflows, floods, fires and seismic activity.

The secondary consequences

Lava deposits in Santorini Island (Greece) (by A. Vassilopoulos, N. Evelpidou).

concerns all the long-term consequences of the volcanic activity for the environment and human activities, such as long-term climate changes, destruction of biotopes and residential areas, increasing rate of desertification, etc. Analytically, the effects of volcanic activity are attributed to:

- *Lava flows:* They take place when the magma reaches the surface and runs over the crater covering the volcano's sides. Lava flows are the typical products of volcanic activity and are characterised by high or low velocity and low or high viscosity, respectively. Most of the lavas move quite slowly, so people have the chance of reacting and protecting themselves. Many methods have been adopted for the diversion of the lava flows such as bombings, hydraulic freezing and construction of barrier-walls and canals.

- *Pyroclastic activity: This is associated with* magma with high silicon concentration. During this activity, all kinds of suspended pyroclastic material, varying from volcanic dust to ash, eject from the volcanic tube to the atmosphere. These explosions are quite intense and of high velocity which possibly exceeds the speed of sound, thus the materials can be transferred to great distances and cover hundreds or even thousands square kilometers. Pyroclastic activity may have direct consequences for the environment and specifically for fauna and flora, while major catastrophes can take place in residential areas and for infrastructure.

- *Poisonous gases:* A number of gases such as CO_2, CO, H_2S, etc are emitted during and between periods of volcanic activity. Usually, these gases are heavier than air, so they remain close to ground, sometimes resulting in numerous deaths. A characteristic example is the Cameroon case on the 21st August 1986, where 2000 people died due to poisonous gases.

- *Caldera explosions:* The caldera explosions are extremely violent and of enormous size, but they are quite rare. An explosion of this kind can eject violently more than 15000 cubic kilometers of pyroclastic material creating a huge caldera-crater with an area of a thousand square kilometers. Calderas explosions have been noticed even in the recent geological past.

- *Mudflows:* They are caused by the saturation by water of a great volume of volcanic dust and other volcanic products, which results in the creation of a massflow of significant velocity. The material volume can be up to millions cubic meters and the velocity of 100 kilometers per hour. The mudflows are dangerous and they have immediate effects in the environment.

- *Fires:* Fires are caused around the volcano due to high temperatures, which can reach hundreds of degrees Kelvin.

- *Seismic activity:* Seismic activity is often a precursor of the following volcanic activity and usually accompanies the volcanic explosions.

Residual landforms consisting of pyroclastic material. Santorini (Greece) (by A. Vassilopoulos, N. Evelpidou).

main volcanic landforms

ACIDIC ROCKS

Rocks whose SiO_2 level is higher than 63%. They originate from viscous magmas, more or less saturated. They form flows, domes or needles. They are most often found in the ejections of pyroclastics.

Nisyros-Greece (by K. Kyriakopoulos)

BASALTS

The most well known volcanic rocks. They are mafic rocks, microlithic, of the gabbro family, with feldspar plagioclases, calciferous without quartz.

Santorini-Greece
(by K. Kyriakopoulos)

BASIC ROCKS

Rocks whose SiO_2 level is lower than 52%. They originate from very fluid magmas, effused at a very high temperatures, 1,100 to 1,200°C. Spreading widely,

they may cover thousands of km².

Orthris-Greece (by K. Pavlopoulos)

ULTRABASIC SERIES

Rocks without quartz or feldspars in their mineral composition. Their SiO_2 level is lower than 45%.

Sousaki-Greece (by K. Kyriakopoulos)

BOMB VOLCANIS

They are fusiform lava pieces formed during rapid cooling. During their ejection, they rotate in the air, and due to their semi-fluid state they get the typical form of a bomb.

Santorini-Greece (by A. Vassilopoulos, N. Evelpidou)

CALDERA

Crater whose size is measured in kilometres, created by an eruption or the collapse of the volcano's central section.

Santorini-Greece (by A. Vassilopoulos, N. Evelpidou)

CRATER - MAAR

Eruption crater directly opened in the substratum.

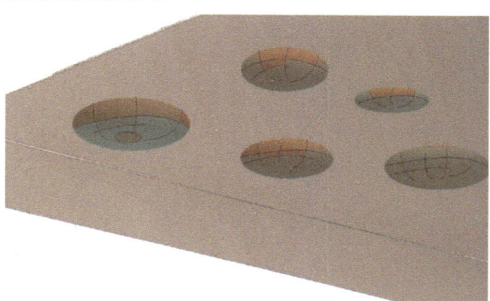

HOMOGENISED CRATER

Cavity of width varying from tens to hundreds of metres corresponding to the opening of one or more volcanoes.

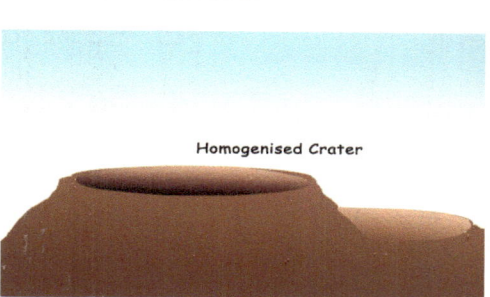

Homogenised Crater

CONE CRATER

Crater opened on the top of a cone of lava or pyroclastites.

cone crater

DYKE

Volcanic landform devived from a lava stream. Due to differential erosion processes in the area, the dyke remains in the form of a wall.

Sithonia Peninsula-Greece
(by K. Kyriakopoulos)

FUMAROLES, VOLCANIC AREA WITH VAPOURS

Emissions of overheated vapours that are hydrogenated with sulphur and fill the air with sulphur deposits.

Flegrean fields-Italy
(by K. Kyriakopoulos)

GRANITE

The most common underlying rock of the continental masses . Granite's mineral composition is: quartz, alkaline feldspars, plagioclase, small amounts of dark minerals like biotite, hornblende and less often pyroxenes and tourmaline. It also contains insignificant, light-coloured, complementary minerals, such as muscovite, lithionite, apatite etc. Depending on the kind and amount of the secondary components granite is distinguished as: biotitic, muscovitic, two-mica granite, hornblende bearing granite, biotitic–hordblende bearing granite, hypersthenic etc. Their structure is characterised as granular panallotriomorphic to hypidiomorphic. Granites occur in many forms and particularly as batholiths and stocks.

also in irregular plutonic masses.

Naxos-Greece (by A. Vassilopoulos, N. Evelpidou)

GEYSER

A source that produces water and vapours periodically, with peripheral deposits formed mainly from SiO_2. Geysers occur in areas with recent volcanic activity. Their water is meteoric. It percolates to great depths, where it gets heated to very high temperatures. Significant geysers occur in Iceland, New Zealand and the Yellowstone Park (USA).

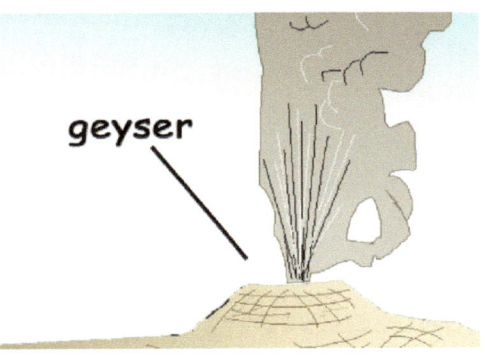

geyser

Cyclades-Greece (by Th. Godelitsas)

GRANODIORITE

A very common rock whose mineral composition is: plagioclase, K-feldspar, quartz, biotite, hornblende and insignificant complementary minerals. Its structure is usually iso-mesogranular, hypidiomorphic with gradations to allotriomorphic. It occurs in the form of batholiths, stocks, expanded veins and beds and

LACOLITE

Lava lenses intruding an area's formations, having cooled in a subhorizontal level.

LAVAS

Lenticular lavas with visible flow characteristics.

Aigina-Greece (by K. Kyriakopoulos)

DOME OF LAVA

Cylindrical formation with steep and convex border slopes, with protrusions of non-differentiated lava and high viscosity.

Methana-Greece (by K. Kyriakopoulos)

PRISMATIC FORMS OF LAVAS

Prismatic landforms created on a superficial lava mass because of its rapid cooling.

Bolsena-Italy (by K. Kyriakopoulos)

OPHIOLITES

This is a complex of basic (gabbro) and ultrabasic

(peridotite) magmatic rocks, of the products of their metamorphism, and also of serpentinites (green rocks) and basalts. They are considered as indicators of oceanic crust from oceans that have now disappeared. The name ophiolite (ophis=greek for snake, lithos=greek for rock) occurs from the typical green colour and the scaly appearance of serpentinites (from the latin word serpens = snake). The ophiolithic complexes are characterised by a specific succession, which from bottom to top consists of: a) tectonites b) the chamber of rocks c) the microgabbro veins and finally, d) the surface volcanic rocks, usually in the form of pillow lavas, with alternations of pelagic sediments. An incomplete or disrupted ophiolithic succession is an indication of tectonic events, mainly during its displacement towards the continental rims.

Lamia-Greece (by Th. Godelitsas)

TYPICAL WAVY SURFACES

The top surface of a lava flow, characterised by wrinkles transversal to the flow direction.

Etna-Sicily (by K. Kyriakopoulos)

THERMAL SPRING

A spring of high temperature water, due to the volcanic or magmatic origin of the water table, that sometimes gushes under gas pressure. Thermal springs occur mainly where expanded faults that can reach great depths exist. More than one spring may form along a fault, and then so called "thermal spring lines" are formed. The common existence of hydrogen sulphide in thermal springs is due to the reduction of sulphuric salts, principally of gypsum, and to the carbonic acid that is continuously released from the depths of the earth, mainly in areas with plutonic intrusions.

Thermopylea-Greece
(by K. Pavlopoulos)

VOLCANIC BRECCIA

Course-grained fragments (>2mm) lithified with a cement of tephra or lapillae. Breccias contain fragments of lava and resistant rocks.

Cyclades-Greece (by Th. Godelitsas)

LAHAR-VOLCANO OF SEDIMENTS

A cone, oblate on its edges and closed by to the accumulation of liquid deposits by argillaceous rocks that originate from gas explosions.

Milos-Greece (by K. Kyriakopoulos)

Pyroclastic material in Santorini Island-Greece (by E. Efraimiadou)

Weathered granodiorite formations in Naxos Island (Greece) (by A. Vassilopoulos, N. Evelpidou).

Tunisia (by A. Vassilopoulos, N. Evelpidou)

Chapter 8

aeolian environments

aeolian processes

Aeolian transport and deposition

A wind speed known as the "fluid threshold velocity" is necessary for sand transport. This speed is proportional to the size of the sand grains and the relation between them is, in general, positive, which means that the bigger the size of the sand grains the higher the fluid threshold velocity needed for the transport. For very fine fractions, such as silt and clay, with high cohesion between the grains, this relationship reverses because of significant resistance in the movement. When the grains start to move, transport can be effected by drifting, i.e. the grains move along the surface. If the wind speed reaches a certain level, the sand grains can lifted by the wind and are transported in suspension. The combination of these two processes is the most typical mode of aeolian sand transport, known as "saltation". During this process, the sand grains, after their initial movement by airflows above the fluid threshold velocity, are transported by the wind for a short distance and eventually fall to the ground. The sand grains that bounce, come in contact

Residual landform in Tunisia due to aeolian erosion processes (by A. Vassilopoulos, N. Evelpidou).

with other grains that hold kinetic energy, resulting in the lowering of the fluid threshold velocity of the sand grains. This reduced velocity is known as the "impact threshold velocity". Consequently, it is clear that the sand transport can be sustained even in low wind speeds, after its initiation.

Sand deposition requires the reduction of the wind speed. For instance, in the coastal zone this reduction takes place on the lee side of obstacles like woods, shells, bushes etc. The aggregation of the sand due to the wind forms a dune which is characterized by a downwind side of gentle slope and a lee side of a steeper slope. The sand inside the dunes is usually deposited in a specific structure, known as "cross- bedding". It is characterised by the presence of crosscutting sand layers with small and large angles of inclination, that represent older downwind and lee sides of the dune.

Sand and wind interaction

The wind speed over a sand surface is reduced due to the friction, just like the water that flows in a river. Wind currents, that are prerequisites for the transport of the sand material from the beach inland, are created by the differential heating between the land and the sea. The starting point for the aggregation of the sand and the formation of the dunes is the coastal vegetation. Coastal dunes are favoured in beaches of gentle slope, with a high tide range, because the sand depositions exposed to the wind are extensive.

During the first stages of their formation the dunes are called "embryonic" and are usually destroyed by the wind. They are called "stable" when covered by vegetation.

The substantial difference between coastal sand dunes and other coastal landforms is that their formation depends on the wind rather than the water movement.

Coastal sand dunes lie above the high tide level of the coastal zone and usually represent the limit of marine action on the coast. They can extend over the land up to 10 km from the coastline and often act as a coastal barrier that protects the lower coastal areas from the sea.

Coastal dunes differ from other types of sand dunes. Despite the fact that the basic formation process for both coastal and desert dunes is the aeolian transport of the sand, latter they have a totally different morphology. The process that distinguishes the desert from the coastal dunes is the interaction between the wind and the vegetation that takes place in the coastal dunes and not in the desert dunes. However, in dry regions where the coastal zone carries no vegetation, coastal and desert dunes have similar morphological characteristics.

Coastal sand dunes are aeolian landforms that appear mainly in dry, semidry and hot climates, rather than in tropical and subtropical areas where their formation is limited by the dense vegetation, the low wind speeds and the high humidity of the sand.

The area where coastal dunes can form may also include longshore sand bars, parallel to the coastline and separated by longshore troughs. There are dune systems with highly complicated morphology, like for example when the dune ridges are vertical or form acute angles with the coastline.

The coastal dune ridges can range from 1 or 2 m to 20 or 30 m in height, while their gradient is usually sharp towards the downwind side and more gentle on their lee side, in contrast to desert dunes. They have flat or wavy tops. Sometimes they have low troughs, with no vegetation, known as "blow-outs".

The appropriate conditions for the formation of coastal sand dunes include:

- An extended inland area of the coastal zone that is able to host aeolian sand depositions as for example on coasts of low relief without cliffs.
- The appropriate wind regime. For sand transport, strong winds blowing in a stable direction are required. The dunes are usually formed along the coasts or areas that are often influenced by storms. The height of a dune is determined by the wind speed, so the highest dunes are formed in areas exposed to strongest winds.
- Large quantities of sand of right grain size. Dunes are always formed by the transfer of the sand by the aeolian processes. Additionally, the stability and the development of the dunes require a constant supply of sand. Well developed dunes have significant dimensions and are usually close to the sources of sediment supply, such as a river mouth where their source material is transported from the drainage basin to the shore.
- Vegetation helps in the concentration and the stability of the sand. Dunes without any vegetation, like those in dry desert conditions, can be also be formed in coastal environments, as for example the areas where the desert meets the sea, that is to say the desert coasts (i.e. Namimbia, Africa). Coastal sand dunes are usually developed when the transport rate of the sand is high and exceeds the vegetation development rate. Dunes that are free of vegetation are called "free dunes" and are sensitive to changes of wind direction. These dunes are often lortogonal to the prevailing wind direction. Additionally, dunes may accord with the vegetation development. The presence of vegetation on the surface of the dunes helps with their stabilization, since it eliminates the loss of sand material and the migration to the inland. The impeded dunes, arrested by vegetation, are orientated more aligned to the source of the sand than to the direction of the wind.

A well developed aeolian sand dune with ripple marks in Tunisia (by A. Vassilopoulos, N. Evelpidou).

main aeolian landforms

AEOLIAN DEPOSITION

Deposition which generally consists of material transported by wind.

Naxos-Greece (by A. Vassilopoulos, N. Evelpidou)

AEOLIAN EROSION

Wind transportation of a load of small sized grains (<2mm) that originate from a ground surface that is dry and without cohesion.

Tunisia (by A. Vassilopoulos, N. Evelpidou)

AEOLIAN SURFACE

An eroded surface, with stripes, smoothed or scratched by wind-transported sand.

Medano del Coro National Park-Venezuela (by C. Centeri)

ARROW OF SAND

It is an sand structure ranging in size from some centimetres up to some meters. It is formed by the wind behind a topographic obstacle.

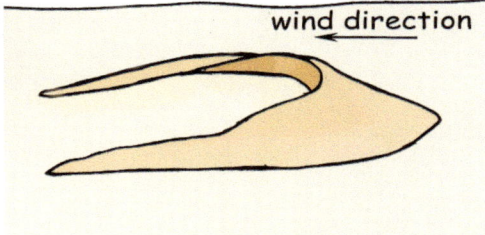

wind direction

BARKHANE

Crescent dunes larger than 10m, with a gentle inclination towards the wind's direction, that are concave downwind with a steep slope. It is easily affected by the air currents and forms humps. A bisymmetrical barkhane, with humps of different size, is called an "elb" (alab on the plural).

Tunisia (by A. Vassilopoulos, N. Evelpidou)

DIRECTION OF THE DRASTIC WIND

A drastic wind can directly lift and transport sand.

Aeolian erosion landforms in Tunisia (by A. Vassilopoulos, N. Evelpidou).

Aeolian sand dunes in Tunisia (by A. Vassilopoulos, N. Evelpidou).

Tunisia (by A. Vassilopoulos, N. Evelpidou)

DUNES

Accumulations of sand due to wind activity.

Tunisia (by A. Vassilopoulos, N. Evelpidou)

DUNE FIELDS

A group of dunes characterised by the same or similar geometry.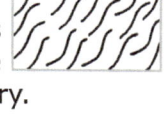

DUNES LENGTHWISE

Dunes aligned, or in levels, following the direction of the wind.

PARABOLIC DUNES

Dunes of crescent form with their concave  side turned towards the wind's direction. It is a form of aeolian erosion–accumulation: the material is extracted in large quantities and accumulates in the direction of the air current in a rosary arrangement. It is a common effect in coastal areas.

DUNES SIDE TO SIDE

Dunes arrayed and inclined towards the  wind's direction.

DUNES IN A WEB

Network of dunes formed in two directions.

NEBKA

Dune which ranges from some centimetres up to  some meters and is created behind a bush in the direction of the wind.

EXPORT

Aeolian transportation of very fine-grained 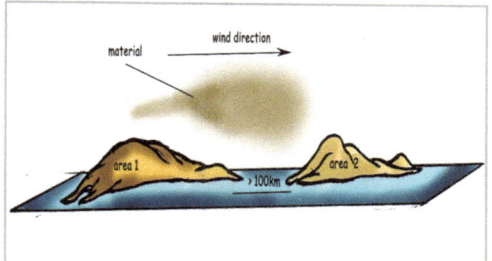 material to a great distance. Continental air currents can carry fine-grained material (<50μm) to a great height and to a distance of thousands of kilometres.

Meteora - Greece (by K. Pavlopoulos)

Chapter 9

surface landforms

BAHADA

An expanded alluvial surface consisting of a series of neighbouring alluvial fans, which have been joined together through time. This expanded alluvial surface can spread for many kilometres beyond the front of a mountainous block.

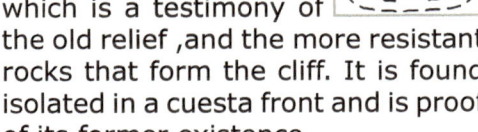

Tatra mountains-Slovakia (by C. Centeri)

BUTTE

A hill with a flattened top, which is a testimony of the old relief ,and the more resistant rocks that form the cliff. It is found isolated in a cuesta front and is proof of its former existence.

The Wild Yeliou Park-Taiwan (by S. Liakopoulos)

BASIN

A depression on the relief (sink) visible on the ground surface. It may originate from different phenomena (e.g. tectonism, glaciation, aeolian erosion, karstic phenomena, hydrogeochemical phenomena).

CALCIC CRUST

A compact crust of calciferous matrix, produced by chemical or biological intrusion/redistribution of the carbonate, on the interior or the surface of a pedologic profile, superficial formation or rock. The transportation of carbonates requires abundant water. The calcic crusts of arid or semi-arid areas in the modern era are generally lithified and have been formed during the Quaternary.

FYROM (by K. Pavlopoulos)

BLOCKFIELDS OR FELSENMEERE

Great accumulations of blocks found on mountain summits and semipolar areas, created by the fragmentation of large rock blocks due to hoarfrost.

Lavrion-Greece (by N. Tsoukalas)

CLASTITES

The result of mechanical weathering of rocks. These formations have the same composition as bedrock.

Samos-Greece (by A. Vassilopoulos, N. Evelpidou)

COLLUVIUM

A generally heterogeneous formation that consists of transported material; created by the physicochemical destruction (i.e. due to high inclination) and the accumulation of material on the foot of the slope. It occurs under the influence of gravity forces, weathering and soilflow.

Vouraikos-Greece (by A. Vassilopoulos, N. Evelpidou)

CORE STONES

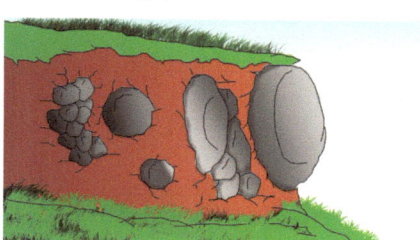

Products of spheroid weathering not connected to bedrock after the weathering process.

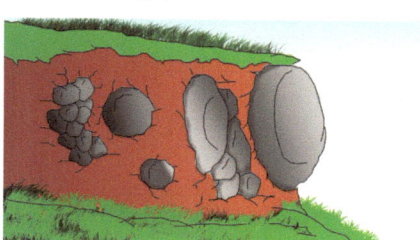

CORNICHE

The rim formed on the edge of a plateau, or a planation surface, that borders a steep slope.

EARTH HUMMOCKS

Small spherical hills that consist of fine material and are covered by thick vegetation.

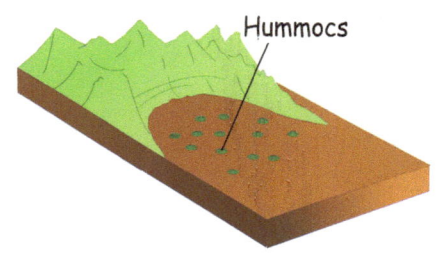

Hummocs

HOMOCLINIC LANDFORMS

CUESTA

It is a monoclinic structure located in areas that consist of two different rocks with the resistant rock lying over one that is more easily eroded. In every case the relief consists of an anticlinic front and a reverse cataclinic side. The term cuesta corresponds to the French term côte de Lorraine. This term is used to avoid possible confusion with the term côte, which refers to coastal geomorphology.

HOMOCLINIC RIDGE

A homoclinic form characterised by inclinations ranging from 10° to 30°.

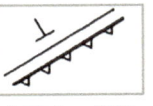

ACUTE RIDGE

A homoclinic form characterised by inclinations ranging from 30° to 70°.

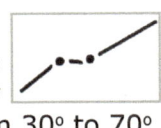

SPAR

A homoclinic form characterised by inclinations higher than 70°.

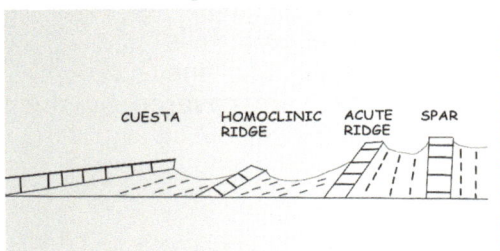

calcareous ones of the karstic area of Istria (former Yugoslavia). Despite the fact that these landforms are mainly developed on sedimentary rocks, their development has also been observed on magmatic rocks, as in the area around the cistern rock of Henry Mountains in Utah. Due to high inclinations, hogbacks do not retreat easily, as is also the case with cuesta landforms. In the areas where rocks are preserved in slopes or in dome elevations, differential erosion, by the drainage network, smoothens the acuminations, an effect known as "flatiron".

Naxos-Greece (by A. Vassilopoulos, N. Evelpidou)

HOG-BACK

The term Hogback is used to describe a long narrow ridge, or series of hills that structurally consist of sedimentary formations characterised by high inclination values. When the inclination of the layers is higher than 50%, an almost symmetric hogback landform is created. In cases of inclination lower than 50%, the inclination of the ridge depends on that of the sediment layers. The ridges with inclination less than 40% are sometimes called homoclinic backs. The hogbacks are developed, usually, in successions of soft and hard sedimentary rocks and their creation is favoured by the outwash and erosion in semi-arid climates. The most known hogbacks are the

MESAS

Form of sedimentary rocks with horizontal or sub-horizontal inclination covered by harder and more resistant.

METEORIC CRATER

A round cavity created by a meteorite fall.

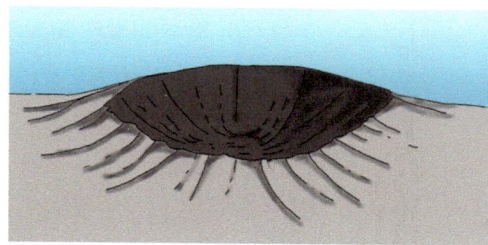

MONOCLINIC SLOPE, FACE

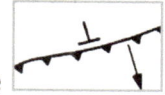

A high inclination slope consisting of a resistant rock that lies over a less resistant one, and forms the front of a monoclinic relief.

Erymanthos-Greece (by N. Tsoukalas)

PENEPLAIN

Surface characterised by very low topographic inclinations formed due to the erosion that the whole area has sustained. The formation of a peneplain is the last stage of the erosion cycle of the relief. Chronologically, the deposits of a peneplain are always considered to be older than the ones that cover them and posterior to the most recent layers that have been eroded.

Blue Mountain-Australia
(by K. Pavlopoulos)

PLANATION SURFACE

Planation surfaces are located in mountainous areas and are characterised by very smooth relief. They are created due to rocks' weathering and the erosion of the relief in an environment of mild tectonism. Planation surfaces are very important for the following reasons: a) They represent periods of tectonic tranquillity and humid-warm climate during the development of mountain masses. b) The present location of the erosion surfaces demonstrates the incidence of intense faulting and uplifts of mountain masses. c) The surfaces of erosion located in a higher altitude than others are chronologically older. The flat sections of the planation surfaces are destroyed due to the processes of the relief's development and particularly due to exogenous processes.

PLATEAU/INSELBERG

A flat trapezoid surface, located higher than its neighbouring environment. The transition from a typical plateau to the lower sections is usually characterised by a steep relief. A plateau can be the elevated zone between faults, the top of a broad fold, or a tectonically elevated peneplain. There are various classifications that generally distinguish plateaux

179

as intramontane, foothill and continental.

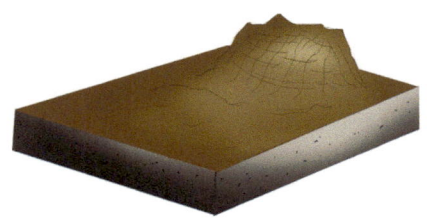

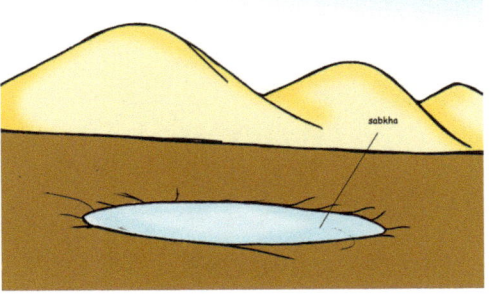

PLAYA

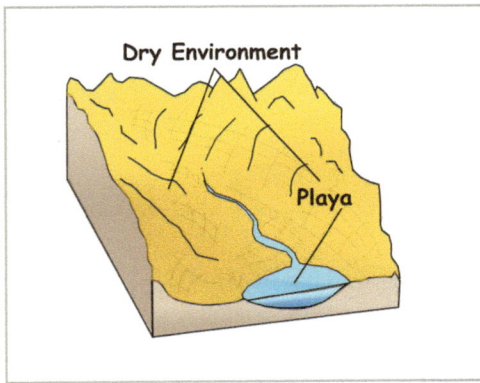

A flat and smooth plain containing significant percentage of saline components. It contains argillaceous material and is located in the centre of an already drained basin. After intense rainfall, it is possible for the Playa plain to be covered with water and a temporary shallow lake, rich in fine-grained material (Playa lake) may possibly form.

Dry Environment

Playa

SLOPE

Inclined surfaces of rocks, soils or even loose sediments of various inclinations, higher than 5°. Slopes may occur by processes of weathering and erosion, by tectonic movements, in more rare occasions by deposition, or generally by a succession of the above mentioned processes.

N. Peloponnesus-Greece (by A. Vassilopoulos, N. Evelpidou)

STONE CIRCLES

An isolated form of the stonenet.

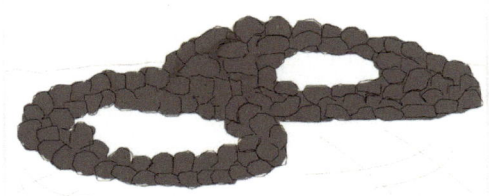

SABKHA

Lowering of the ground surface with a flat floor that has a salty composition (salt or gypsum). It is an area flooded in rainfall periods, in desert and dry areas with high evaporation.

STONE NETS

A mosaic surface that consists of clay, silt and gravel in the centre and coarser material on the perimeter.

STONE RIVERS

Mass movements of blockfield material under the effect of gravity.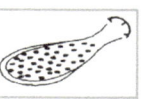

Erymanthos-Greece (by N. Tsoukalas)

STONE STRIPES

Parallel stripes of stones and fine-grained 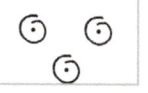 material that occur on slopes with steep inclinations.

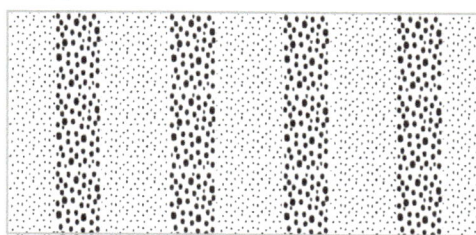

TAFONI

The term Tafoni is given to landforms 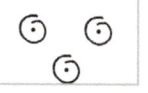 found in the area with the same name in Corsica. They are forms of celliform weathering found principally in crystaline rocks of

medium to large grain size, but can also be found in other rocks, such as sandstones, limestones and schists. The dimensions of these cavities range from a few centimetres up to several meters. Cavities of diameter of 20-30 centimetres are sometimes called "tafoni miniatures".

Naxos-Greece (by A. Vassilopoulos, N. Evelpidou)

TOR

These landforms of spheroid weathering are huge naked rock blocks, located on the ground and surrounded by a multitude of smaller rocks. The term has been initially used to describe the granite mountain block in Dartmoor, SW England; however, its meaning has been expanded to include similar structures in various rocks, in a multitude of climatic environments. In Africa, the term Kopje, or Koppie in the local dialect, is principally used for similar landforms. The height of these spheroid weathering landforms, rarely exceeds 17 meters, and is usually much lower. Tors can be found in an important variety of topographic environments, such as mountain summits, watersheds, smooth banks, slopes, and sometimes even valley floors. They can be classified as the "skyline tors" that take up the most elevated places in the area, and "sub-skyline tors", which can be found on the banks of valleys and within sinks. Tors are principally characteristic of thick-grained porfyritic granites. They are rarely found in schists, but they have been recorded in sedimentary rocks, such as quartzites and feldspathic sandstones.

Tinos-Greece (by D. Leonidopoulou)

WATERSHED

The limit of a drainage basin, defining the accumulation area for water, through a network of underground or superficial drainage channels.

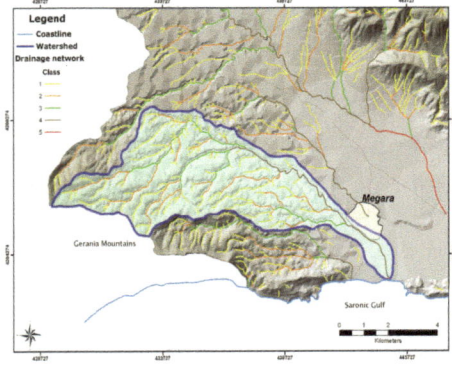

Hogback landforms in Santorini Island (Greece) (by A. Vassilopoulos, N. Evelpidou).

Sporades - Greece (by A. Vassilopoulos, N. Evelpidou)

Chapter 10

topography, lithology and tectonics

TOPOGRAPHY

ALTITUDE

Expressed in metres and centimetres, measured from sea level. On maps this is given the value 0.

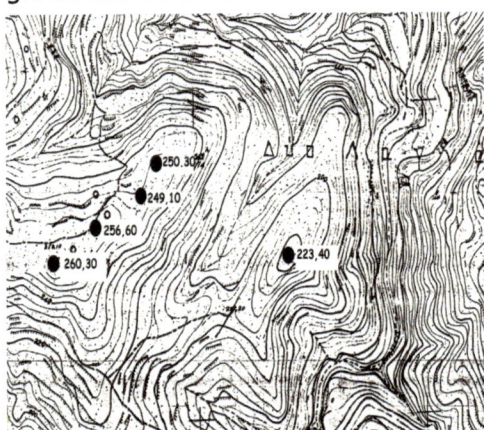

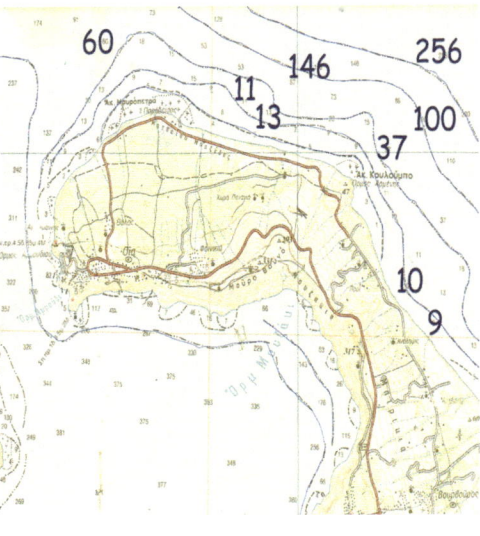

CONTOUR

A curve connecting the points of equal altitude. The constant vertical distance between two successive contours is called contour interval and expresses their altitudinal difference.

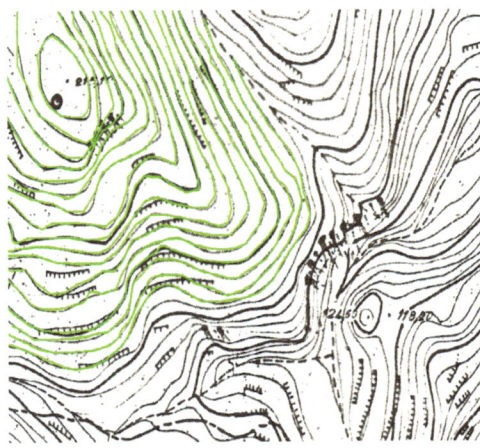

ISOBATHS

Curves connecting points of equal depth.

INCLINATION

The distance between two contours expresses the relief's inclination. As the distance gets smaller, the relief gets steeper, when it gets bigger, the relief is more gentle.

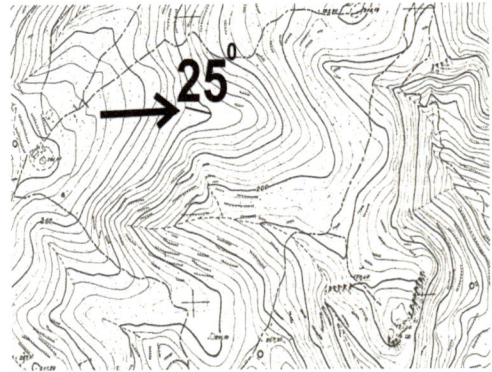

INCLINATION < 10°

A relief with inclination <10°; characterised by diffuse water flow. It is also expressed by the tangent of the angle formed by the relief's surface and the horizontal plane. The value of the tangent, multiplied by 100, corresponds to the % inclination of the relief.

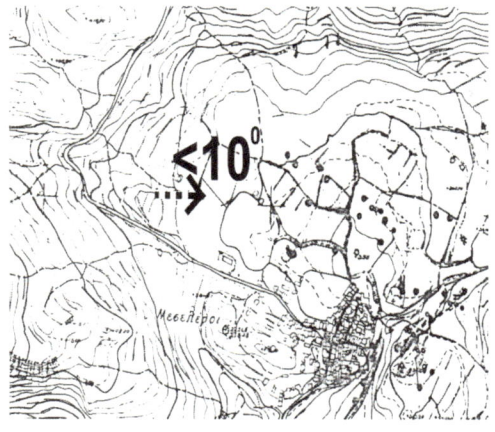

INCLINATION 10°-30°

A relief with inclination varying from 10° to 30°; characterised by channels and rills. At the same time underground water circulation and soilfluxion takes place.

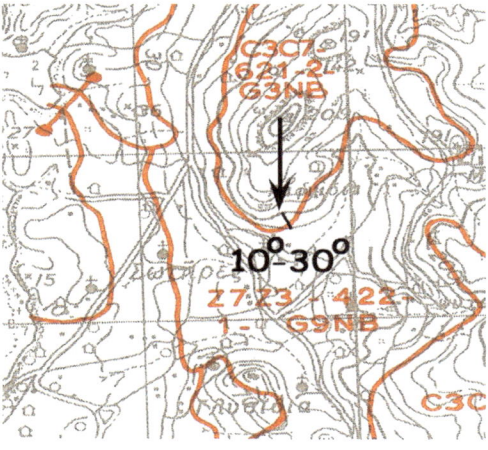

INCLINATION OF A LAYER , A SURFACE OR A DIS- CONTINUITY

The inclination of a stratigraphic layer or a discontinuity surface is the angle formed between the surface or the layer and the horizontal plane. This inclination must not be confused with the topographic inclination, which is the angle formed by the topographic surface and the horizontal plane.

China (by S. Liakopoulos)

INCLINATION > 30°

In inclinations higher than 30°, gravity plays a predominant role. In these areas, and under certain conditions that depend on the rock quality, the soil composition, the discontinuities, the soil water etc, there is a significant chance for phenomena such as landslides, collapses, mudflows, soilcreeps etc, to occur.

SEDIMENTARY FORMATIONS

DOLOMITES

A sedimentary rock containing more than 15% of magnesium carbonate. As a mineral it is crystallised in the triangular crystallisation system. Its colour is rose or grey, it has a glassy shine, a specific gravity of 2.85 – 2.95 and hardness of 3.5 - 4. It is affected by a dilute solution of HCl. It is used as building and decorative material and in the production of MgO for fire-resistant materials.

Attica-Greece (by I. Matiatos)

EVAPORITES

An accumulation of salt minerals, as a result of the evaporation of waters containing them. The waters in arid areas can be of lacustrine or lagoon phase. Their morphological and geotectonic role depends mainly on the way they were deposited. The three principal evaporitic minerals are halite, gypsum and anhydrite.

W. Greece (by Th. Godelitsas)

GYPSUM

Calcium sulphate generally exists in nature in the form of anhydrite ($CaSO_4$), or in the hydrated form ($CaSO_4*2H_2O$) of gypsum. These two forms are in general correlated with calcite and dolomite. The transformation of anhydride into gypsum (e.g. when the atmospheric air is very humid) is accompanied by a volume increase, capable to dismember neighbouring rocks. Gypsum occurs in bands or lenses within limestones and clays, or in stones within sands (desert rose). Other appearances are observed in thick, fibrous or fine-grained blocks, within the epicontinental sedimentary successions, or even uncovered by erosion in the centre of diapiric folds. Finally, it is a fragile and soluble formation that, when conditions are favourable (as in a humid climate), leads to the formation of karstic landforms.

Cyprus (by N. Tsoukalas)

FLYSCH

A heterogeneous formation consisting of sediments varying in grain size and thicknesses, such as sandstones, roundstones , marls and silts. The flysch originates from terrigenous material provided by the friction between a submerging plate and another plate, and is deposited within the trench basins of orogenetic arcs. The flysch, silt content, creates slippery surfaces during winter, leading to the landsliding of the overlying deposits.

Argolida-Greece (by I. Matiatos)

GRANULAR PHASES

Transferred rocks, consisting of grains varying in size, together with silicate or calcite grains.

GRAVEL

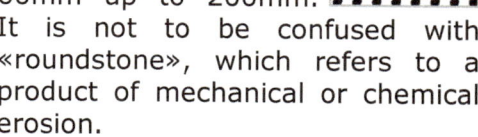

Sediment grains from 60mm up to 200mm. It is not to be confused with «roundstone», which refers to a product of mechanical or chemical erosion.

Attica-Greece (by K. Pavlopoulos)

SANDS

Sediment grains whose diameter varies between 2mm and 50mm. Depending on the nature of the prevailing components it may be distinguished as quartz, feldsparic, limy, oolithic and organic sand (the last originating from shells). Sand is also distinguished as sea sand, aeolian sand and alluvial sand, according its origin.

Naxos-Greece (by A. Vassilopoulos, N. Evelpidou)

SILTS

Grains' dimension ranges from 50 – 2μm. Silts are characterised by high capillary retention. Silts break up only under arid conditions. Their superficial break-up facilitates their transportation by the wind.

Attica-Greece (by K. Pavlopoulos)

CLAY

Clay is characterised by high plasticity and contributes to the creation of argillaceous rocks with thin layering (crystallites <2μm) linked by a thin layer of water assimilated in the salt components' ions. The porosity and the thin layer of water (called pore–crystal) render clay impermeable to free water. In arid conditions, clay maintains its coherence (contraction fissures or absorption ion of water of the atmospheric precipitates). If the percentage of water originating from capillary effects increases, the fissures close again, the rock surpasses its "plasticity limit" and transforms into an unstable and easily deformable rock. If the quantity of water increases further, it reaches the "liquid limit", at which clay freely outflows . These limits are called "Atterberg limits". The impermeability of clay to water renders it geomorphologically quite important.

Cyclades-Greece (by Th. Godelitsas)

LIMESTONE

Rock that contains more than 90% $CaCO_3$. The rest of the material that comprises it may be argillaceous, ferric or magnesitic. Limestones are distinguished, according to their origin in rocks of: a. detrital, b. chemical, c. organic, origin. Limestones weather and break up more easily than other rocks. Rain water, through chemical decomposition (solution) causes karstification, thus creating various landforms, such as caves, dolines, poljes and uvalas.

Attica-Greece (by Th. Godelitsas)

MARL

Fine-grained formation, soft, loose or friable. A mixture of 35% clay 65% calcite. Its main feature is plasticity deriving from its content of clay (>35%).

Evia-Greece (by I. Matiatos)

MOLASSE

Term generally referring to a formation that consists of terrigenous material provided by the mountains' weathering during orogenesis and deposited within the fore-trench of the orogenetic arc. Its cohesiveness varies from low to very high. It is a mixture of sandstones, conglomerates and silts by deposition of fresh or salt waters. Cartographically, the molasse is depicted either by its constitutive elements or in a combination of a neutral colour lineage on an also neutral colour basis.

Karditsa-Greece (by D. Theocharis)

SOIL PROFILES

Soil sections defined according to the rules of pedology. These profile-sections are registered in stations and are represented on maps by a cross, accompanied by the depth indication in centimetres.

• 0,25

Lublin-Poland (by A. Vassilopoulos, N. Evelpidou)

TRAVERTINE

A calcite deposit. It may be located at the outlet of a karst spring, on the borders of a waterstream, or on the brinks of a waterfall. It originates from the chemical settling of calcium carbonate in supersaturated waters.

Erymanthos-Greece (by N. Tsoukalas)

METAMORPHIC ROCKS

GNEISS

These rocks are characterised by granular texture, with medium or larger grains. Foliation may be apparent as layering, for example in the arrangement of the dark coloured layerings (Fe-Mg minerals) in alternation with light coloured layerings (quartz and feldspars).

Attica-Greece (by A. Vassilopoulos, N. Evelpidou)

MARBLES AND SIPOLINES

These rocks result from the metamorphism of limestones or dolomites. They have a crystalline form, relatively good lustre and white colour (marble) or various tints (sipolines).

Attica-Greece (by Th. Godelitsas)

191

TECTONICS

ANTICLINE

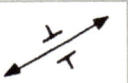

Convex fold in the higher section of a stratigraphic layer with diverging legs. The older sedimentary layers are located in the interior of the fold. An anticlinic fold whose axis length is slightly smaller or equal to its total expanse is called a brachyanticline.

The Wild Yeliou Park-Taiwan (by S. Liakopoulos)

Naxos-Greece (by A. Vassilopoulos, N. Evelpidou)

ANTICLINE AXIS

The axis of a fold, on either side of which the stratigraphic layers are dipping off the axis; thus the layers are sinking in relation to the axis that is the higher part of the fold.

DIACLASE

Surfaces along which the rocks have been fragmented. They are characterised by a small displacement vertical to their surface, and by no or little displacement of the two separated segments parallel to their surface. The diaclase's limits vary from a few millimetres up to some centimetres. When the diaclases occur in abundance in a rock and have the same geometrical features, they form a group of diaclases.

FAULT

Discontinuity in a rocky, rigid mass, with large relative displacement. This may have components parallel and vertical to the surface of the two segments that are split by the fault. The displacement varies from a few centimetres to several kilometres and can influence very large pieces of the Earth's crust. When activated, earthquakes are generated, which is why their scientific study is of great interest. There are many classes of faults, each based on different criteria. For example:

a. Depending on the relative displacement of the sections, they are distinguished in dip-slip and strike slip. In dip-slip faults the relative movement between the two segments is vertical and they may be further distinguished into regular, reverse or thrust. In strike-slip faults the relative movement between the segments is horizontal and can be sinistral or dextral

b. Depending on the inclination of their surface they are distinguished in narrow or wide angle faults

c. Depending on the relation between the layers' and fault's aspect they are distinguished as concordant and opposite,

d. Depending on the correlation of

the slide vector with the trend of the fault, they are distinguished as slide faults by inclination or by trend, or even of side sliding.

The amount of a fault's displacement is usually measurable, in relation to the displacement of the two segments' stratigraphic layers. This displacement is always measured parallel to the movement's direction and is known as the fault's "throw".

Corinth-Greece (by A. Vassilopoulos, N. Evelpidou)

HYPOTHETICAL FAULT

A fault is considered hypothetical, when it derives from the study and interpretation of the lithology, topography or the drainage network.

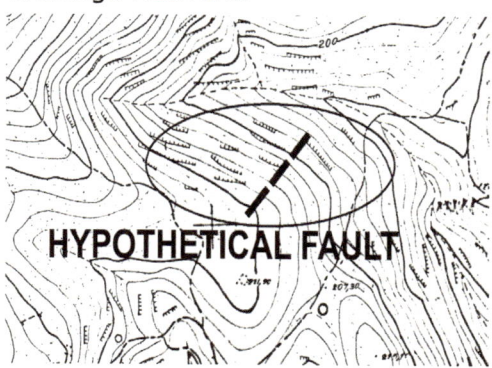

HYPOTHETICAL FAULT

FISSURES

Longitudinal notches, of small depth in the

substratum, due to the friction caused by ice, the aeolian erosion and the widening due to dissolution.

Peloponnesus (by E. Efraimiadou)

GRABBEN

Lowered land section, whose borders are two neighbouring faults dipping towards the lowered section.

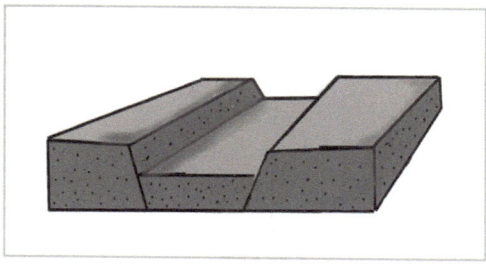

HORST

Elevated land section whose borders are two neighbouring faults dipping away from the elevated section.

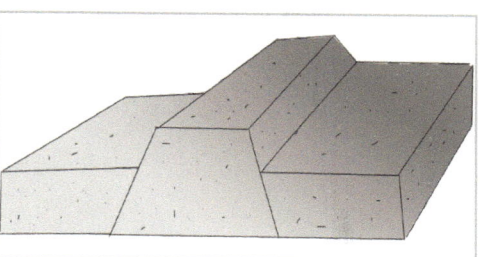

FAULT SCARP (DIRECT OR PRIMARY)

A topographic altitudinal difference (D) between an elevated and a lowered piece of ground, directly created as a result of the fault's tectonic movement. The scarp can be «active», «inactive», or «dissimilar» (formed during the phases of successive activation or tectonic tranquillity). The extent of the altitudinal difference is defined by the fault's throw.

Attica-Greece (by D. Theocharis)

SYNCLINE

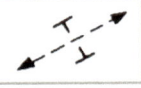

Concave fold with converging legs. The older sediment layers are located in the exterior part of the fold. A synclinic fold whose axis length is slightly bigger or equal to its total expanse is called brachysyncline.

Kalavryta-Greece (by I. Matiatos)

SYNCLINE AXIS

The axis of a fold, on either side of which the stratigraphic layers are dipping towards the axis; thus the axis is the lower part of the fold.

A scarp of fault in Samos Island (Greece) (by C. Centeri).

Montmorency falls - Canada (by N. Tsoukalas)

Chapter 11

geomorphological
mapping
(case studies)

Case study 1: Geomorphological study of the Oinois river (North Attica-Greece)

The Oinois (or Charadros) River is located in northeast Attica (Greece). The total main riverbed length is about 31Km, while the drainage basin covers an area of 177,2Km². It is bounded to the west by the ridge of Mt. Parnitha and to the south by Mt. Pentelikon. The watershed height to the north is about 500m, where it seperates the Oinois drainage basin from several smaller drainage networks to the south that cross six fault zones of E-W and NW-SE directions before they terminate at the Euboic Gulf. The Oinois River starts with an E-W flow direction in the upper part of the drainage basin (a mountainous area), changes, a little, to a NE direction at the midpoint of its course and finally discharges at Marathonas bay (southern Euboic Gulf).

The discharge area of the river is characterised by an alluvial fan which constitutes Marathonas coastal plain, widely known for the famous battle of Marathonas between the Greeks and the Persians in 490 B.C. The plain, whose long axis is aligned NE-SW, is divided in two sections by the Oinois river. West of the alluvial fan, lies a marshy area that was drained a few decades ago; in the eastern part we find Marathonas marsh which is seperated from the sea by a sand barrier and is characterized by the formation of low relief coastal sand

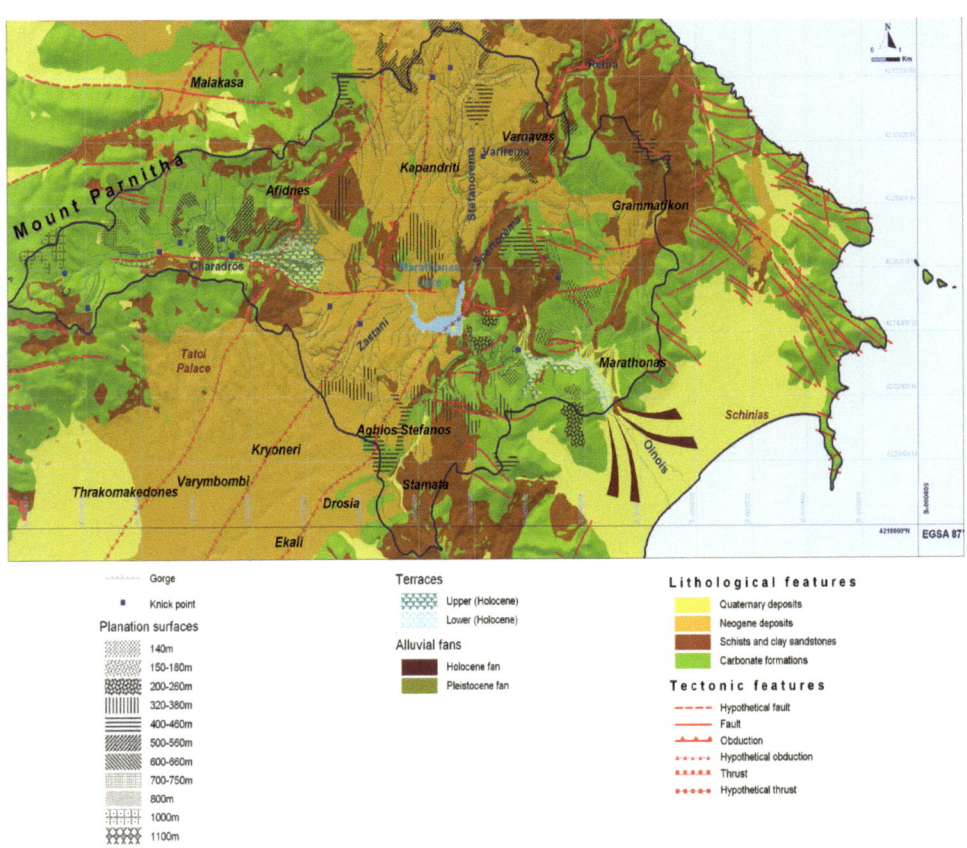

dunes, stabilised by the vegetation.

About 12Km above the river's estuary lies the Marathonas dam, constructed in 1929, whose reservoir has been used for the water supply of the Athens basin for a long time.

The study of relatively small drainage basins in areas where the precipitation height is rather low (about 500mm or less), offers important information about their morphotectonic evolution. The Oinois river drainage basin is a typical example and in order to examine its geomorphological evolution during the Quaternary it was essential to map all the landforms found in the basin, to specify the spatial distribution of the morphometric parameters of the drainage network and to correlate both of them with the tectonic features and the lithological characteristics of the drainage basin. Additionally, it was important to map the geomorphological characteristics of the alluvial fan in the estuary and to detect temporal changes of the coastal zone. The thorough examination of these changes led to the conclusion that, to a large extent, they may be attributed to human activity.

Methodology

The geomorphological study of the Oinois River drainage basin included the quantitative geomorphological analysis of the hydrographic network and geomorphological mapping in the field. Also, the geomorphological evolution of the alluvial fan in the Oinois river estuary required the geomorphological mapping of the coastal zone and the detection of its temporal changes due to both physical procedures and human activity.

In order to carry out the measurements, a Geographic Information System (G.I.S.) was designed and developed. For each of the following morphometric parameters: a) hydrographic frequency, b) hydrographic density, c) slope inclination and d) circularity, the mean values per class were firstly calculated and then plotted on variability diagrams.

Additionally, it was instructive to create a cross section of Oinois river main riverbed from topographic maps of scale 1/25.000 and to estimate the inclination together with the rest of the morphological characteristics in different sections of the riverbed.

- The preferred scale for the geomorphological mapping of the drainage basin was 1/25.000, while photomaps, received in 1986 by the Hellenic Military Geographical Service, were used for the illustration of the landforms.

- Geomorphological mapping of the coastline of Oinois River deltaic fan was carried out on topographic maps at 1/5.000 scale, provided by the Hellenic Military Geographical Service, while the coastline temporal changes were estimated with the help of old maps and several photomaps dating from 1938 to 1988.

Geomorphological mapping

The geomorphological characteristics of the Oinois River drainage basin are depicted in the geomorphological map at 1/25.000 scale.

The planation surfaces are located at different heights from 140m to 1100m. The ones with the lowest heights (140m and 150-180m) are found in marble formations SW and W of the Marathonas region, while those with heights 320-380m are in breccia conglomerates of the Kapandriti and clay schists of the Afidnes Unit to the N and NW of the Marathonas reservoir, respectively. The planation surfaces with heights 400-460m lie in marble formations of the Aghios Stefanos region and to the NE area of the reservoir lake, while those of 500-560m have been formed wholly in breccia conglomerates of the northern part of the drainage basin. It is important to notice that the planation surfaces are found at increasingly higher altitudes (600-660m, 700-750m, 800m, 1000m and 1100m), as we move towards the western part of the drainage basin. These surfaces have been formed in limestones of the Pelagonian Unit, some of the highest, of which have undergone dissolution processes producing karstic landforms.

In the Afidnes area, two terraces are found along the main riverbed of Oinois River. The upper one reaches 2m in height while the lower one 1-1,5m. Their formation has taken place in the Holocene and is the result of the slow and continuous tectonic uplift of the upper part of the drainage basin occurring from the Middle Pleistocene until today, which also led to the deposition of the alluvial fan during the Upper Pleistocene. Terraces are also found in the upstream parts of the alluvial fan of the estuary. At the debouchment of the gorge, below the Marathonas dam, lie two terraces, the lower 1m and the upper 2m in height, the formation of which has taken place during the Holocene. Additionally, terraces of low height are located across several tributaries of the northern drainage network (Stefanorema, Paliomothi).

Significant downcutting erosion processes have been noticed along the main riverbed of the Oinois River upstream in the Afidnes area and in the whole drainage network of Kapandriti region which has developed in breccia conglomerates of the Upper Miocene. The formation of the gorge in the lower part of the river is attributed to headward erosion processes. In the upper part of Afidnes region the aforementioned gorge is the result of in depth erosion processes due to the tectonic uplift of the area since the Middle Pleistocene, while in the upper part of the river the gorge's formation has been facilitated by the evolution of the river across a tectonic discontinuity in the E-W direction.

Geomorphology of the coastal alluvial fan

The mean inclination of the coastal alluvial fan of the Oinois River was estimated at 1% (the peak of the fan reaches 20m in height and is 2Km from the coastline). The erosion processes of the main river, at that point, are quite intense attaining 5m in depth. When the Oinois River enters the alluvial fan, it separates in a western and an eastern stream, known as the Sehri river and the Kainourgio river, respectively. The main riverbed of the Sehri river has been inactive for several centuries and it divides in smaller branches as it reaches the coast. This riverbed does not exceed 2m in depth, and covered by soil and vegetation. As shown on the topographic map of Curtius – Kaupert of 1989, mosaics and ruins from the Roman period were found in the ancient riverbed of the Sehri river, proving its inactiveness.

In the photomaps of 1938 and 1945, an older riverbed was detected west of the present ones. This riverbed passes through the northern part of Marathonas tomb and discharges into the sea in a SE direction.

The present riverbed of the Kainourgio river follows a parallel route to the Sehri river. The river Kainourgio has been inactive in recent decades, as it is concluded from the extensive samples of sand, the presence of wastes and the artificial debris deposition observed in the riverbed which minimises its width to 2m in the estuary. The main reason for the interruption of the flow of the river Kainourgio is the construction and operation of the dam in the main riverbed of the Oinois River.

Geomorphological characteristics of the coastal zone

The coastline of the Oinois river alluvial fan has an almost linear shape, except the area in the eastern section of the Kainourgio river estuary. There, the coastline bends inland. This is attributed to the granular differences of the coastal sediments.

The inclination of the alluvial fan along the coastal zone is small (less than 20%). The coastal sediments in the estuary of Kainourgio river consist of coarse grained material, mainly conglomerates and gravels of a diameter which infrequently exceeds 20cm. In the eastern part, these sediments include mainly sand mixed with conglomerates and gravels, while in the western part the proportion of the sand increases, as we reach the Sehri river estuary.

The SE winds that blow in the region represent only 18% of the total wind frequency. When the wind reaches force 7 Beaufort, the wave height can exceed 2m, causing an east direction coastal transport of all the fine grained material from the area around the Kainourgio river estuary towards the sand barrier located to the east. At the same time, there is a secondary current, less significant, that transports sediments to the west.

Along the whole coastline, apart from the area around the Kainourgio river estuary, the observed coastal sand dunes are old, stabilised, covered by vegetation and their height hardly exceeds 1,5m. Currently, part of these dunes has eroded, due to the coastline regression caused by the coastal processes.

At least seven older river estuaries have been recognised in the wider area of the recently banked up Kainourgio river estuary. The comparison of a series of photomaps of 1938 and 1988, led to the conclusion that in the estuary of the Kainourgio river a regression of the coastline has taken place. This is estimated at over 100m, which corresponds to a regression rate of 2m per year, over the last 50 years. This regression can be attributed to the presence of the Marathonas dam, constructed in 1929. The dam operation caused significant changes in the physical processes, resulting in the deposition of river sediments inside the reservoir behind the dam and the decrease of the river flow and the material transport to the estuary. It is also important to mention that the anthropogenic activities in the riverbed of the Kainourgio river, such as the extended sand extraction, have made the riverbed inactive. Thus, there is no flow in the riverbed even when extreme phenomena, such as strong storms, occur in the lower part of Oinois river, between the dam and the estuary.

The comparison of the topographic map at 1/25.000 scale and the photomaps of 1938 showed that there are no significant coastline changes in this period.

The coastline regression of the last 50 years is not expected to continue at the same rate. However, it is almost certain that it will be taking place for several years in the future, until the old riverbed of the Kainourgio river, lying almost parallel to the coastline near its estuary, is destroyed by the sea.

One more reason to expect this coastline regression in Marathonas plain, in the near future, is anticipated global sea level rise. According to recent studies dealing with the global climate change (because of the greenhouse effect), the air temperature of the planet will increase about 2ºC, which will lead to a mean sea level rise of about 49cm in the next 100 years (I.P.C.C. 2001). This probable rise will have a serious impact on the rate of coastal processes and eventually the regression of the coastline.

Human activities in the coastal zone

Until 1938, there were no constructions near the coastline of the alluvial fan. The coastal plain was cultivated to a great extent, and it was divided in several smaller, mainly linear, plots. The plots in the first 300m of the coastal zone lie vertically to the coastline, while those inland are parallel to the coastline. It is obvious that this different plots arrangement is caused by coastline advance due to the deposition of transported material, not only from the old Sehri riverbed, but also from the later Kainourgio riverbed.

The Kainourgio river has remained inactive for several years after 1949, the year when the Marathonas dam was constructed. The city of Athens was less inhabited at that time, so its water needs were limited.

Between 1945 and 1960, the first developments, 100m in width, appeared in the part of the coastal zone between the Sehri and Kainourgio rivers. There were eight developments, most of which were used for summer holiday habitation. At the same time, an increase of the coastline regression rate was observed.

During the 1960's, an intensive residential development took place in the coastal zone, which was expressed by the construction of at least 20 houses. The coastal erosion processes were still intense.

In the 1970's, a coastal asphalt road running to the west near the Sehri river was constructed. At the end of this decade, an old dirt road which traverses the banked up estuary of the Kainourgio river started to erode, while, at the same time, a soil breakwater was built at the western part of the estuary.

In 1984, the inhabitants increased to 48 and another two small breakwaters had already been constructed the year before, both sides of the Kainourgio river. The eastern one was earthen and resisted erosion only for a few years, while the western one was built from concrete, about 10m in width. It is estimated that the coastline regression rate, at that period, was approximately 1m/year lower than the rates observed later decades.

In 1988, the number of inhabitants reached approximately 65. The concrete breakwater started to erode due to coastal processes.

These erosion processes continue. Along the coastline, the locals inhabitants are trying to prevent sea water intrusion by building the coast with material or by building concrete walls, but with no success.

Case study 2: Geomorphological study of the Attica basin (Greece)

The Attica basin can be separated in the following four geomorphological units: Imitos, Penteli, Parnitha and Egaleo Mountains.

The eastern and southern part of Mt. Imitos has a discontinuous linear shape. The southern part is divided from the northern part and it seems to have moved eastward to a lower altitude. Planation surfaces inside the drainage basin of Ilissos River (NW section of Mt. Imitos) are observed in 400-440m, 680-700m and 920-980m and are correlated, according to their height and lithology, with the ones from Mt. Penteli.

In the western section of the mountain, along the riverbed of the Ilissos river, downcutting erosion processes are quite intense, as it is evidenced by the presence of a 1,5Km length V shape valley extending from 380-520m in height. It is believed that the "separation" of the mountain took place along a NNE-SSW axis located a few hundred meters west of the mountain top, but no tectonic element that proves this theory has yet been found. However, debris cones of relatively small inclinations are detected on some mountain slopes. In general, the northern part of the mountain's western slopes show constant inclinations around 23-25%; in the central part the inclinations reach 31%; while in the south they decline gradually to 4%. High inclinations, over 100% (45°), are observed in the NW part where the heights reach 700m and 900m.

Mt. Penteli lies in the SE part of Attica basin. It is a symmetrical mountain ridge with steep slopes, especially in the SE section. In general, the mean inclinations of the slopes vary between 7% and 24%. In the SW part there are shallow valleys with high inlcinations. Planation surfaces in the basin are observed at 720-800m, 460-500m height.

Mt. Parnitha is the biggest and highest of the four mountains and covers the NNE part of Attica basin. In altitudes between 800m and 1000m, its surface is highly eroded. There, the V shape valleys have undergone downcutting erosion processes, as in the three V shape gorges in the SW part of the mountain. The latter are found between 500m and 1200m in height, while their length is 2,5Km, 3Km, 3,2Km respectively as we move from West to the East. Their depth varies between 100m and 300m. The formation is probably due to tectonic uplift which has taken place in the region, during the

203

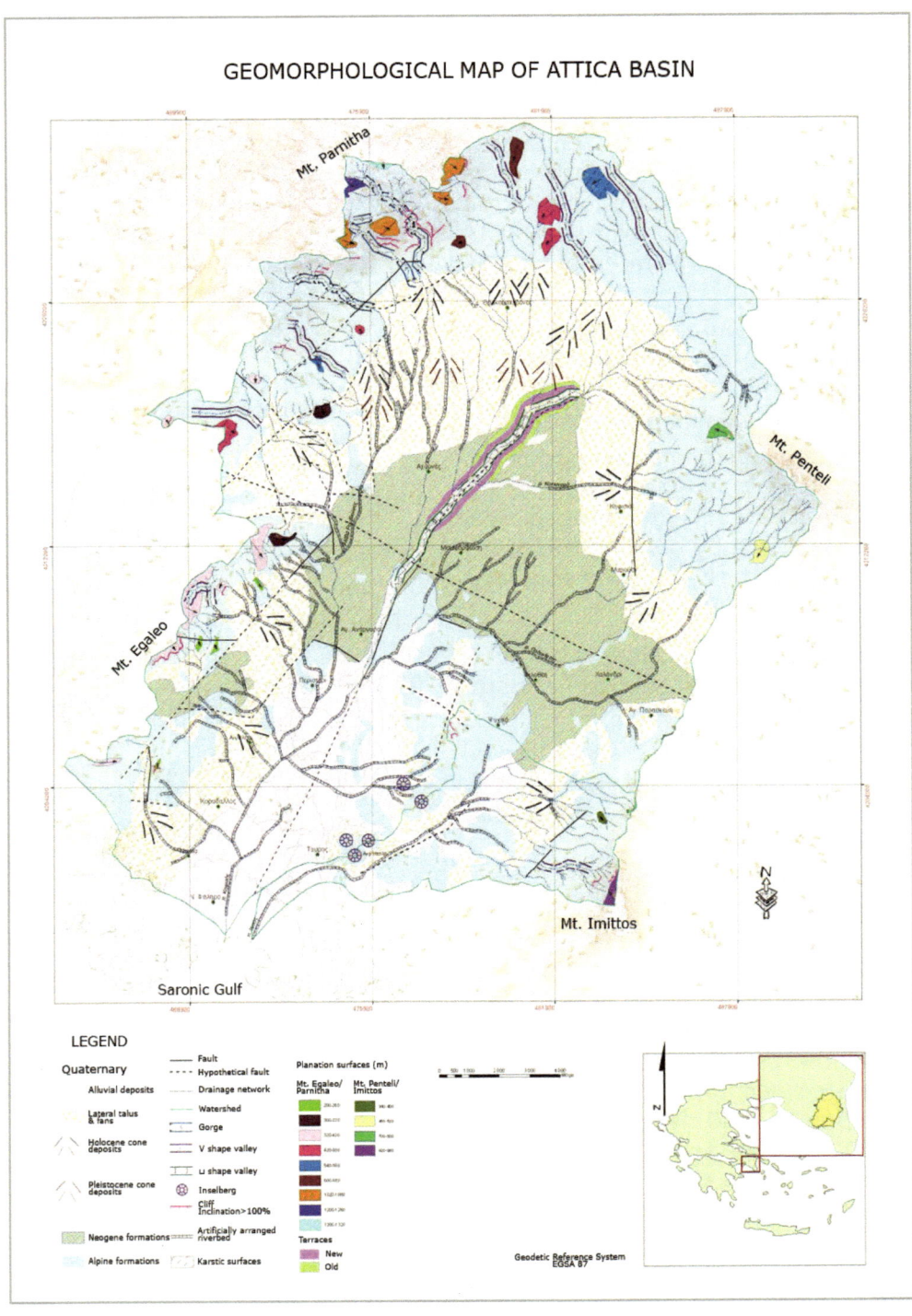

GEOMORPHOLOGICAL MAP OF ATTICA BASIN

Upper Miocene (23 m. y. ago). The sections. It reaches 75% in the inclination of the mountain slopes SW, 65% in the central part and differs in the various mountain 22-28% in the NE. Moreover, high inclinations (over 100%) are not rare. Finally, the planation surfaces of low inclination (<10%) are found at different heights such as 300-320m, 360-440m, 420-520m, 540-560m, 600-680m, 700-780m, 1020-1080m, 1200-1600m, 1300-1320m. These surfaces show an inclination towards the interior of the drainage basin and are characterised by karstic landforms.

Mt. Egalaio has a linear top that exceeds 300m in height. It is divided in two sections, with the southern one being higher than the northern. Planation surfaces are observed at 200-260m and 340-420m height, while high inclinations (>100%) feature in the southern mountain area, adjacent to the planation surfaces. The southern slope of the mountain varies significantly in inclination between 8% in the north and 13% in the south. Finally, a V shape valley of 1,4Km length, due to downcutting erosion, is also observed in this area.

The Athens plain

The foothills, between the mountainous areas and Athens plain, are characterised by planation surfaces of small inclination. They are covered by fluvial deposits, while in the lowlands, north of Mt. Egalaio and SW of Mt. Imitos, we find salty deposits of Neogene age. Alluvial cones of Holocene age are dominant, though we can find a few of Pleistocene age lying at low altitudes, in the southern part of Mt. Parnitha.

The Athens plain extends from the foothills to the coast and the altitude in the mountain base does not exceed 400m. The plain is characterised by small inclinations varying between 1,5-6,5%, excluding the hills with much higher inclinations. The largest part of the plain lies in front of Mt. Parnitha and Mt. Egalaio. Its inclination in the NNE-SSW direction is approximately 2%. At 100-400m height we can find intensively eroded valleys up to 10m in depth. The latter are attributed to the climatic change of the last Glacial period, when the sea level was 120m lower than the present, and also to the tectonic uplift of Mt. Parnitha. The lowest parts of these valleys are covered by modern fluvial deposits. This deposition in the Athens plain has been probably continuous at least during the Quaternary. Significantly, we see that the northern part of the region has U shaped valleys due to in-depth erosion on the upper course of Kifissos River. In the same area several terraces are also present.

The hills of Athens plain

The hills have a NE-SW orientation and separate the Athens basin into two parts, eastern and western. If we traverse the basin from NE to SW, we will observe four main hills: Tourkovounia (323m), Lycabetus (265m), Acropolis (142m) and Filopappou (161m). These four hills consist of rocky formations, remnants of a previous relief which characterised the last formation of the Athens plain, and now can be considered to be inselbergs.

It is almost certain that the sea during the Neogene reached the Acropolis region probably surrounding it for a short time. Pliocene deposits have been found to the west of Lycabetus hill at 120m height. Deposits of the same age are more frequent in the west of the northern part of Mt. Imitos, where

205

many alluvial cones and fans are also present. Finally, in Tourkovounia hill clay formations have been observed in faults and diaclases of limestones of Pleistocene age. Secondary hills are also present such as Arios Pagos (115m), Asteroskopeion (104m), Pnyx (109m), Filopappou (147m) and Kolonos (68m).

Case study 3: Geomorphological study of Paros Island (Greece)

Paros Island lies in the Aegean Sea and belongs to Cyclades complex. The island has an ellipsoid shape and covers approximately 196Km². The coast has deep indentations in the NE, where Naousa bay is found, and also in Paros bay (Paroikia) in the NW. The relief in most of the island is steep with rock formations; other parts are covered by sand deposits. The topography is mountainous in the middle of the island, with a highest altitude of 771m at Profitis Ilias summit; it is flat in the coastal zone. The largest plains on the region are: Naousa, Marmara, Dryos and Pounta. The vegetation in the island is relatively poor.

During a geomorphological study of Paros Island, a large number of inclination measurements (176 in total, one value every 500m) were carried out starting at 500m from the coastline and moving inland, so as to determine the spatial distribution of the topographic inclinations along the coastal zone of the island. The inclination values were grouped and classified in the following four main categories of the same range:

- 24,7 – 32,2 % (class A)
- 17 – 24,7 % (class B)
- 9,3 – 17 % (class C)
- 1,6 – 9,3 % (class D)

The highest inclinations are observed in the NW part of the island, between cape Vorino and cape Maistros, except for one measurement of this category in the SE part of the island, south of Kefalos bay. The values of class B occur in the following zones:

- In the NE part of the island, between the western side of Plastiras bay and the northern part of cape Fanos.
- In Fanos bay (southern Paros)
- North of Kefalos bay (one measurement)

Category C includes the following areas:

- NE of Plastiras bay up to cape Vorino
- From cape Maistros up to Paroikia bay
- South of Platia Ammos bay and Latzeris bay
- In eastern Paros, in the wider area of cape Ntamoulis.
- South of Kefalos bay (Marpissa area)
- In the southern part of the island from cape Mavros Cavos up to cape Pyrgos

Unlike categories A, B, and C which feature in only a small part of the coastal zone, category D (inclinations between 1,6 % and 9,3 %) dominates in the island.

In order to study the spatial distribution of the inclination values, we proceeded to overlay of the thematic map of inclination, the lithological map and the geological map. The geology data were obtained from a geological map of I.G.M.E. (Institution of Geology and Mineral Explorarion) of 1996, while the lithological information resulted from the grouping of geological formations with similar lithological characteristics. The combination of these maps showed that the highest inclinations are observed almost exclusively in gneiss schists and especially in the gneiss of the northern part of western Paros. One measurement, taken in molasse formations between Kefalos bay and Piso Livadi, is out of the range of A class. The inclinations of B class are found in gneiss schists of the northern part of western Paros, in carbonate formations of cape Fanos (southern Paros) and in magmatic formations (granites) of cape Plastira (northern Paros). There was only one measurement of this category carried out in clastic formations (molasse sequence) in Kefalos bay (eastern Paros). As far as the values of group C are concerned, they are distributed in all lithological formations, while those of group D feature mainly in clastic deposits, less frequently in gneiss schists and even less in all the other lithologies.

The planation surfaces in the island of Paros are the result of chemical and mechanic erosion process, varying in altitude and age. They cover an area of approximately 2,74Km², most of which (2,17Km²) concerns carbonate formations. Next in area are gneiss schists, with 0,45Km² coverage, molasse sequence, with 0,10Km², and finally magmatic rocks, with 0,02Km². These planation surfaces are found mainly in the western part of central Paros. During mapping, they were classified according to their altitude in 100m ranges. The most frequently occuring altitude ranges lie between 0 and 400m, most often in the 200-300 m range, and consecutively less often in 100-200m, 300-400m and 0-100m ranges. Their frequency above 400m height is low, though they have been identified above 700m height.

The plains of Paros are mainly of karstic or coastal origin. The former are not so frequently observed, except two plains which lie in marbles of the Marathi unit and cover approximately 1,43Km². The coastal plains, which are formed mostly in alluvial deposits and cover an area of 26,02Km², are observed:

- In the Livadia area (western part of Naousa)
- In the Naousa area (east of Plastiras bay)
- In and around Paroikia
- Extensively in eastern Paros from the northeastern part up to Marmara village
- In Dryos bay
- In the Agairia area

Residual erosion landforms are typical in the plains of eastern and western Paros and lie in carbonate formations. Those in the western part of the island are formed in marbles, while the ones of the eastern part are in limestones of Cretaceous age.

The valleys in the island are mostly of V shape, due to intensive downcutting erosion processes; their length reaches 73,11Km. These landforms have been created in the youth stage of a typical river cycle. Additionally, the U shape valleys have gentle slopes, round bottom and 117,8Km total length. These valleys are representative of the maturity stage of a river cycle.

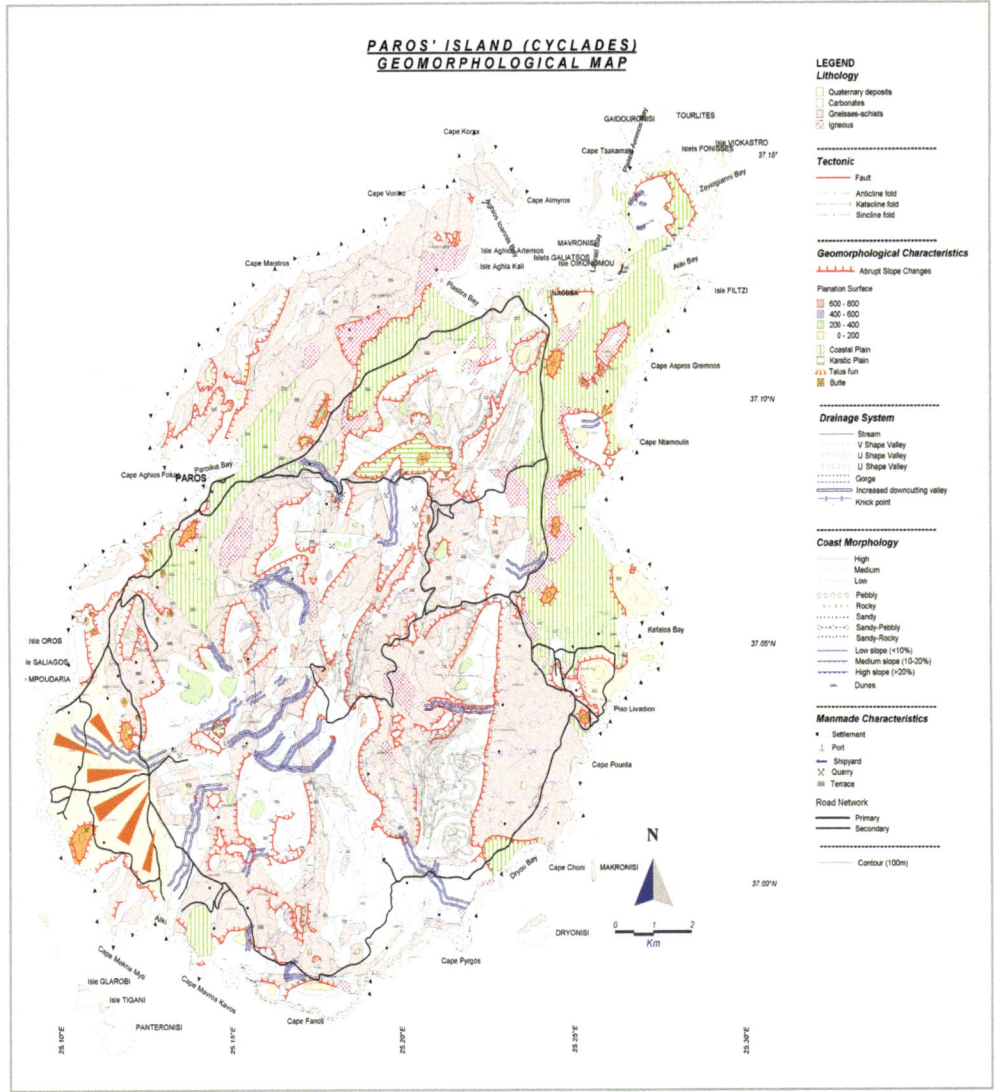

Finally, there are several U shape valleys (representative of the shape of senility of a river cycle) of 47,8Km total length.

The gorges in Paros Island are mostly observed in limestone formations; their creation took place in recent geological time when the climate was more humid than today. Tectonic activity in the island is evidenced by the presence of faults and diaclases. Gorges are found in:

- Southern Paros, east of cape Mavros Kavos
- The SSE part of the island (two gorges in western cape Pyrgos)
- Northern Paros, along the Ksiropotamos river (southern Naousa)
- Western Paros (east of the Paroikia area)

Field work in Paros has also shown that knick points are common in

208

the drainage network; they are attributed to faulting and differences in lithology. The main debris cones in the study area was found in the SW part of Paros (in the Kampos area) and consists of cohesive sandy conglomerate formations which are currently being eroding by the Syrigos river. It covers an area of 9,993Km², while in western Paros (Parasporos stream area) we find smaller debris cones (total area 0.016Km²). In the eastern part of the island (Ampelas area) we find even smaller cones (total area of 0,015Km²).

The sequence of sandy conglomerates of the large debris cone consists of: an upper series with conglomerates of 3cm mean grain size and a lower series with grain size between 45cm and 10cm.

Case study 4: Geomorphological study of Southern Attica (Greece)

Southern Attica, is a peninsula, between the southern region of the Southern Euboic and Saronic gulfs and the southern most part of continental Greece.

The geomorphological study of this area, is mainly based on two methods, one being descriptive and the other quantitative geomorphological analysis, that integrate a number of other methods and results. This study, uses the geological maps of Attica, as well as topographic maps, diagrams and aerial photographs taken during several time periods by the HAGS.

Climatic data from the main climate stations of the study area (such as Ag. Marina - Lavrion, Eleokchorion, Peania and Helliniko) were used to trace the characteristics of the area's climate.

Subsequently and during the descriptive geomorphological analysis phase, a geomorphological mapping and analysis of the described geomorphical mass of the plane was carried out. Several landforms were studied such as planation surfaces, buttes, terraces, etc., as well as man made structures. Particle size, X-ray defraction and microchemical analyses were included in the study and petrographic thin sections were performed on samples of calcareous sandstones.

During the quantitative geomorphical analysis, the three main drainage networks of the study areas (Legrena, Anavyssos and Adami-Potami) were examined. They were analyzed according to the Horton Laws, while the texture and shape of their basins were examined according to morphometric factors: density, frequency, slope, relief, circularity, elongation and lemniscate of the drainage basins. Finally, comparison was made of every factor in each separate drainage network and area, as well as the factors that define and interact with them.

The evaluation of the observations and the results that the two types of geomorphological analyses produced, together with an investigation of the evolution of Southern Attica, mainly during the Quaternary, revealed their importance. We believe, that they contribute to the understanding of today's form of the plain and of the evident problems that mainly concern the land use of this area.

Geology

The southernmost region of the Attica peninsula, which includes the study area, has a very complicated geotectonic structure.

In general, according to our current data, South Attica belongs to the intermediate tectono-metamorphic zone of the Pelagonic unit.

Specifically, the pre-Neogenic rocks that appear in this area, are metamorphic and semi-metamorphic formations: Marbles, dolomite marbles, myca schists and phyllites.

In the Southern Attica area, three main lithostratigraphic units can be found:

1. *Lower geotectonic unit of Attica (relatively autochthonous):* A closer inspection of the first and by consequence most ancient geotectonic unit, establishes the certainty, on one hand, that this "autochthonous" system is metamorphic and severely deformed, with its initial structures facing NE-SW and later ones facing NW and SE, and on the other hand that the system is constituted by a large marble mass - often dolomitic - and by mica - amphibolitic schists. Under the schists, basic and ultrabasic metamorphic rocks appear.

2. *Allochthonous unit of the overthrusted phyllitic system:* The upper schists system, which is a separate geotectonic unit, is overthrusted on the relatively autochthonous substratum of Attica. It is a transgressive series of the Jurassic-Cretaceous layers, the overthrusted phyllitic system and the overthrusted Eohellenic covering. There are imbricared phyllites interpolated by marbles and limestones, prasinites (metamorphic basic eruptive rocks), sericitic schists and quarzites. Limestones and marbles which are found inside the phyllites are frequently brown-coloured, as a consequence of their epigenetic metamorphosis and limoniteosis. This fact has been confirmed by X-ray defraction analysis performed on samples from relevant locations in the area of Ano Sounio.

Apart from the growth and expansion of these two main geotectonic units, a granite intrusion of limited growth in the Plaka Lavriou and Palaio-Kamariza areas of the study area was observed. It is mainly granodiorite with porphyritic texture, which constits of feldspars, quartz, biotite, hornblende, magnetite ea.

3. *The sequence of Tertiary and Quaternary* formation that overlays discordantly on the previous units. From the Tertiary period, only the Neogene is observed while the Paleogene is not obvious. The Neogene layers are composed of conglomerates, sandstones, marls, and marly limestones of a total thickness of some tens of meters. They are mostly lacustrine sediments of fresh and brackish water environments, as is concluded from the fossils Melanopsis, Costala, Planorbis applanatus, Helix and Vivipara. Neogene deposits lie unevenly on the substratum and on the covering phyllites and marble system.

Field observations on Neogene formations showed two different stromatographic units characterizing

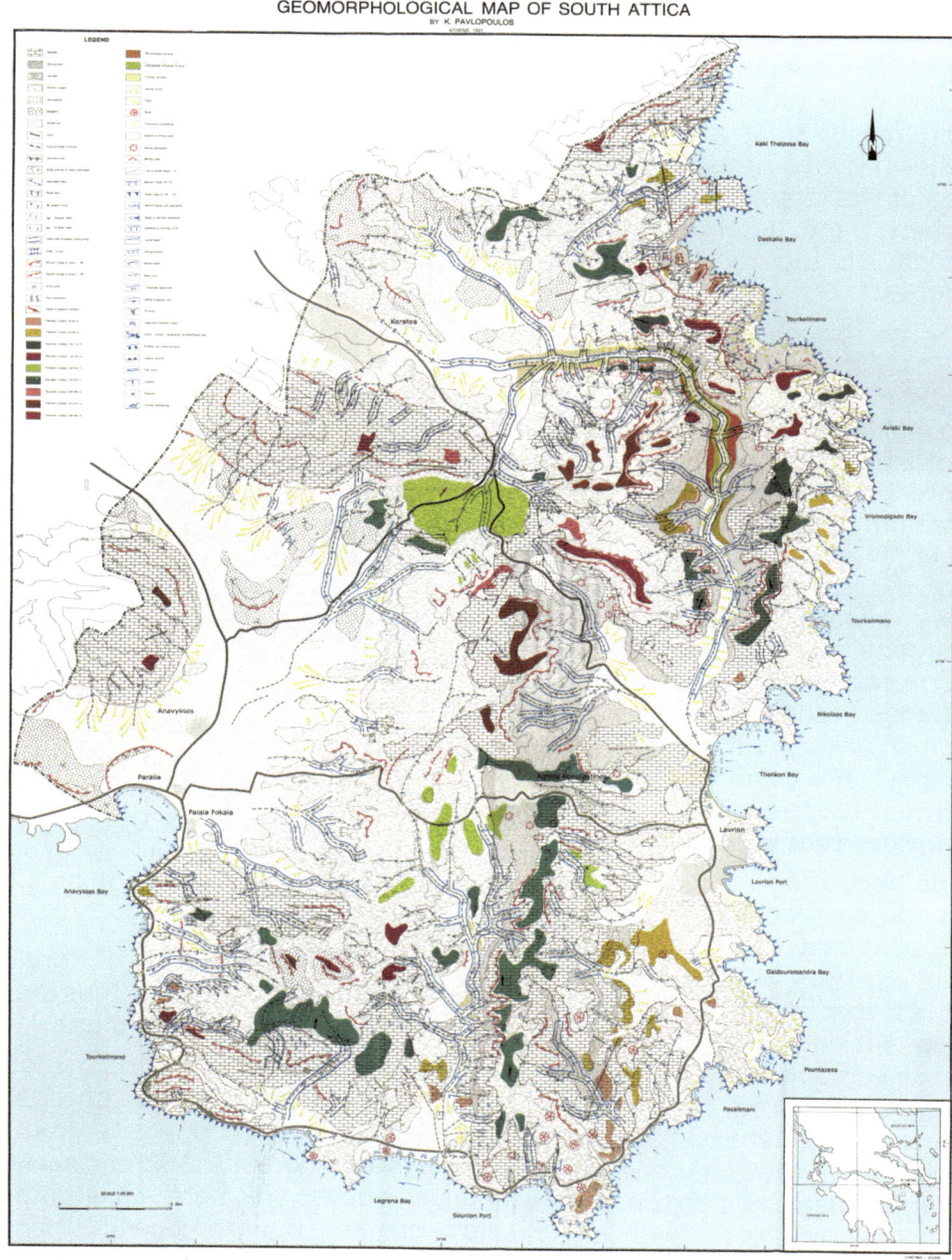

GEOMORPHOLOGICAL MAP OF SOUTH ATTICA
BY K. PAVLOPOULOS

different paleo-environments, known as Ag.Marina-Feriza-Valmas series (brackish phase) and the Anavissos-Kokkinovrakhos unit (terrestrial phase).

• The first unit (Ag. Marina-Feriza-Valmas) is mainly characterised by yellow-green marls with intercalations of conglomerates and sandstones, conglomerates and marly limestones; on top of the unit conglomerates are

211

constituted by pebbles and presenting a characterizing "hole" type weathering. The materials that constitute the conglomerates are mostly weathering products of the crystalic substratum as well as of the Mesozoic limestones and cherts; the latter do not appear today in the rock formations of Attica.

- The second unit (Anavissos - Kokkinovrakhos) includes easily distinguished red-brown conglomerates, which are constituted by pebbles, large sized breccia and marble conglomerates, with small alternation of a red brown clay-marly material. At the base of the red brown conglomerates, we observe large pebbles and breccia of serpentinites, whose size and angular character indicate transportation from a relatively small distance, followed by marly-clay intercalations and very mixed breccia-conglomerates of a fluviotorrential phase.

The fact that in the red unit of breccio-conglomerates there is no evidence of the existence of pebbles from the Ag. Marina unit shows that the first unit is either more ancient than, or contemporary but geographically isolated from the second. The Neogene layers appear to be disturbed and inclining towards NW, proving the influence of newer geologic and tectonic faults in Attica. Normal faults with direction from N30ºW to N60ºW, together with synsedimentary deformations (faults, slides) are typical in both units.

Quaternary formations

These formations overlay discordantly on all older formations and can be distinguished into older and newer deposits.

The former consists of materials from which fluviotorrential terraces, cohesive cones, alluvial fans and cohesive talus slopes are formed. The terraces are found mainly along the Potami valley and consist of relatively cohesive deposits, constituted mainly by marble pebbles and breccia, schists and phyllitic covering ingredients, which are nearby mountain weathering products.

In the relatively newer deposits, we classify the calcareous sandstones which mainly appear in the eastern shores of the study area. This formation has been studied by several researchers and according to Voreades, it represents a marine sediment formed of paleoalluvian, while according to Mistardes it is an aeolian formation deposited during the Quartenary. Negris thinks, that the lower parts of these Quaternary sandstones are marine, while in their higher parts (up to 120m) are of aeolian origin.

These sediments are of great geomorphological interest, thus they were examined in greater depth.

The newer deposits include the low cohesion talus slopes and cones, as well as the recent deposits of the valleys and shores. The old and newer talus slopes are both constituted by coarse, non-homogenous and angular materials; the former have higher cohesion.Alluvial formation include erosion products of mainly metamorphic rocks and appear non-homogenous with loose clay-sandy and gravel ingredients. Recent aeolian deposits, in the form of dunes, are found in parts of the smaller gulfs, such as behind the beaches of Legrena, Kharakas, Anavyssos etc.

A deposit characteristic of the Southeast Attica region is scoriae from ancient as well as contemporary mining. This is found in many places such as: Pasha limani, Pountazeza, Kyprianos, Porto Ennia, Tourkolimano, Viethi, Kamariza, Megala Pefka, Kharakas ea. These formations can be described as recent Holocene deposits caused by anthropogenic activities during historic times. In many cases they resulted in the change of the morphology.

Tectonics

Southeast Attica belongs in the intermediate tectonometamorphic zone of the Pelagonic unit. Its rocks are metamorphic or semi-metamorphic (marbles, dolomitic marbles, mica-schists, phyllites), apart from the Tertiary and Quaternary formations.

In general, in southeast Attica, two main geotectonic units are distinguished:

1. The lower tectonic unit of Attica, which is metamorphic and deformed and considered to be autochthonous.
2. The allochthonous unit of Lavrion tectonically overlain on the autochthonous system of Attica.

The tectonic characteristics of the autochthonous system, have considerable differences from those of the overthrused phyllitic covering. Specifically, on the eastern Lavreotiki, the folding of the marbles and schists of the autochthonous system, mainly lie in a NNE-SSW direction, while the phyllitic system shows a strong folding in the E-W direction. Its' layers have strong inflections and inclinations ($30° - 60°$) accompanied by a number of slides. In western Lavreotiki the anticlinal structures of the autochthonous system have a primary axis in the WNW-ESE direction as well as axes in the N-S direction.

The dividing line of Legrena - Kamariza - Dogani, defines the tectonic difference between east and west Lavreotiki. The tectonic anomaly of Legrena valley is not simply a trap or a faulty fold, but a combined anomaly zone. Initially there was a tectonic rise of the west section until the lower marble was revealed while in the east section the upper marble and the mica schists were preserved. Then the overthrust and placing of the covering above these two sections took place, followed by the submersion of the west side with a pronounced wavy inflection which produced vault folds facing E-W. Many systems of faults exist in SE Attica in other directions.

Climatic conditions

As a whole the Hellenic area is classified, by macro-climate, as Mediterranean. The main characteristics of this climate are the appearance of rain during the cold season of each year and of dryness during the summer. Attica belongs to the sub-tropical zone and the year can be divided in two seasons: The Cold Season (from mid-October to mid-April) and the Warm Season (the remainder og the year).

Attica's climate is a typical Mediterranean climate characterised mainly by a dry and warm summer and a cool and rainy winter. The average air temperature ranges from 16.5 °C to 19 °C with the higher temperatures appearing in south coastal areas and the lower ones in the continental north. The coldest month is January, while July and August are the warmest throughout the year. The average rainfall is 400mm /year and the distribution follows the two main seasons of the Mediterranean climate: rainy (from October to April) with the highest values being observed in December or January, and almost dry during August and July. Snow falls are common in the northern part of Attica, occurring 1-6 days every year, on average. The wind in Attica blows primarily from N-NE directions and secondary from S-SW with an average speed of 5-7 knots.

Relations between relief, climate and hydrography

The dryness of today's climate in conjunction with the recent sea transgression has significantly reduced the solid discharge of several branches of the drainage networks and has helped in the joining together of their valleys in the topographically lower areas. The shaping of the eastern coasts of the study area is strongly influenced by waves. These are created by N and NE winds which appear most frequently and have the highest speeds.

Geomorphological analysis

The purpose of geomorphological analysis is the acquisition of knowledge of the palaeo-environement and its evolution in the study area, as well as finding the best way to use the area's natural potential, without upsetting nature's balance. The geomorphological study and analysis of a certain area are carried out by using one of the following methods:

- Descriptive geomorphological analysis.
- Quantitative geomorphological analysis.

The strong point of the first method is the acquisition of data that depicts a realistic situation, where every landform is unique and distinct from the others, despite the fact that they are alike.

The strong point of the second method concerns the results coming out from measurements of geometric characteristics of the drainage basins. The processing of the data includes analysis of statistical and morphometric factors, which provides us with "guiding" information on the area's geomorphological evolution.

The difficulty and possible weak points of both methods concern the correlation of the data and the geomorphical processes. This is due to the fact that geomorphical processes are "extremely sensitive in the creation conditions" of a certain environment. This means that a slight alteration in the initial genesis conditions (climatic, tectonic, hydrodynamic etc.) can result in the creation of different landforms on the earth's surface. This last idea is expressed by contemporary scientists through the theory of Chaos. Therefore, the evolution of landforms and processes is influenced by the "chaotic" condition existing in all dynamic systems in nature. As a result it seems impossible to absolute forecast the geomorphological evolution of an area.

Geomorphological mapping

By the method of geomorphologic mapping, we can locate, trace and analyze the lanforms which appear in the study area, and therefore we can define the morphogenetic processes that contributed to the shaping of the earth's relief since the Pleistocene. The following are examined:

- The presence of planation surfaces, and the relation between them.
- The types of slopes and their form.
- The shape of the valleys.
- The Quaternary deposits (cones, alluvial deposits, calcareous sandstones and terraces).
- The coastal landforms (Tombolos, beachrocks, littoral longshore drifts, man made structures, dunes ea.).
- The relative sea level changes and their impacts on the coastal environment during the Holocene.
- The recent and ancient human impact on the earth's relief.

Methodology-Equipment

The geomorphological mapping was carried out in two stages:

I. *The coastal geomorphologic mapping*: The mapped coastal zone, includes the region between the area influenced by wave action and the area of a -10m depth approximately. The geomorphological research of the coastal area included:

  - Detailed mapping of the landforms along the coastline in order to define the factors that have affected the shaping of the coastal environment.
  - Submarine research for the

location of the submarine landforms for the acquisition of data related to Holocene sea level changes.

- Sampling in several deposits for laboratory analysis in order to define the physicochemical conditions of their creation and their enviroments.

The coastal mapping was performed during 1988-1989, which means that the presented data (mainly the ones concerning man made structures) is valid up to 1989. For the mapping we used topographic maps of 1:5000 scale provided by the HAGS. The total length of the mapped coastline was approximately 67,5 Km.

II. *Terrestrial geomorphological mapping*: The terrestrial zone, includes the region between the end of the coastal zone and the watershed limit of the drainage basins including the terrestrial zone. The geomorphological research of the terrestrial area was made as following:

- Detailed mapping of the landforms so as to determine the factors that affected the shaping of the present landscape.
- Classification of the landforms and relation of their geomorphical characteristics in order to obtain information about the palaeo-environment during their creation.
- Sampling in several deposits for laboratory analysis. This analysis would define the conditions of deposit creation and formation.

The mapping was performed during 1989-1990 with topographic maps of 1:25000 scale provided by the

Ministry of Environment and Civil Works. The total mapped area was approximately 198 Km². Aerial photographs taken of 1960 provided by the HAGS were also used.

Conclusions

The main conclusions of this study are the following:

1. It confirms the prevailing opinions on the stratigraphy and tectonics of the south Attica area. There is a metamorphic "Attica's substratum" consisting of marbles, mica-schists and dolomites and an overthrusted phyllitic system. Specifically for the Neogene deposits appearing in the study area, two main lithostatigraphic units are distinguished:

   - The Ag.Marina-Feriza-Valmas unit (brackish phase) consisting of yellow-green marles with intercalations by conglomerates and sandstones, conglomerates, marly limestones. On top of the unit, the dominating conglomerates are constituted by pebbles as a product of the "hole" type weathering.

   - The Anavissos-Kokkinovrakhos unit (terrestrial phase) consist of easily distinguished red-brown conglomerates constituted by pebbles, large sized breccia and marble conglomerates, with small alternation of a red brown marly-clay material.

   The Neogene layers, appear to be disturbed and inclining towards the NW, showing the influence of newer geologic and tectonic faults in Attica. Normal faults with direction from N30°W to N60°W, as well as synsedimentary deformations (faults, slides) are typical in both units in the area.

   Therefore, with great caution, we can characterise the Anavissos - Kokkinovrakhos unit as the remains of a molasse formation remain that could possibly be related to the molasse of the Cyclades.

   During the Middle Miocene the south Attica area could have represented a unified palaeo-geographic area, larger in extent than it now is, and with a different morphology. These unified sedimentatary basins were modified during the Higher Miocene-Lower Pleistocene, due to the discontinuous tectonic movements, which resulted in the creation of many small terrestrial basins.

2. According to the geomorphological analysis (quantitative and descriptive) of the study area's drainage networks, the following are concluded:

   - The drainage network of the Potami stream, during the lower Pleistocene, had a flow direction from SW to NE and probably discharged near the area which is today K. Thalassa, Daskalio and Viethi. During the middle Pleistocene, due to discontinuous tectonic movements, the flow direction partly changes, following a N-S direction. At the same time due to tectonic movements and headward erosion processes, a part of the Adami drainage network created a junction with a part of Potami network. This form of "piracy" of the Potami network by the Adami network evolved during the upper

Pleistocene, creating today's complex form of the unified drainage network of Adami-Potami with a sudden change of its flow direction, by 90° approximately, in the area of Viethi.

- From the quantitative analysis of the network, it is concluded that the central part of the drainage network of Adami - Potami is in a rejuvenating evolution stage. This is confirmed by the sudden change in the flow direction, mainly due to tectonic activity, which must have continued during the upper Pleistocene.

- In the drainage network of the Anavyssos stream, a turn towards the western part of the main channel is observed, which can be attributed either to some recent tectonic movements or to an ancient artificial human action.

- The shape of the basin of the drainage network of Legrena, is controlled by the complex Legrena tectonic zone defining the tectonic difference between western and eastern Lavreotiki.

- From the quantitative geomorphological analysis of the three drainage networks, it is concluded that the basins have, in general, a form intermediate between circular and elongated. However, exceptions are the drainage basins of second class streams of the Anavyssos network, which have an almost circular form, as well as those of drainage basins of third class streams of the Legrena network's which tend to be elongated.

- The part of the drainage basins where the greater declinations are observed is controlled by the geology and the tectonics of the marbles and schists of the "substratum" as well as by the tectonic relationship of the latter to the phyllitic system. In these areas, the substratum's marble and schists formations, appear to be recently revealed by erosion processes during the upper Pleistocene-Holocene.

- The observed fluviotorrential terraces, can be distinguished as older cohesive ones and newer ones of low cohesion. Their presence proves the intensive down-cutting erosion processes and their rejuvenating evolution.

3. The planation surfaces, are grouped in four main categories and nine subcategories:

- The planation surfaces from 160-220 m. present, in general, north inclinations (from NW-NE), and show the greater continuity and extent. Their creation period can be considered to lie in the upper Miocene - lower Pliocene.

- The 100-160 m. planation surfaces system is the only one that shows E-SE inclinations; this is contrary to the majority of the surface systems, which have inclinations towards the North. This difference can be attributed to tectonic events which possibly occured during the middle Pleistocene. Their creation possibly took place during the upper Pliocene. This change in their inclination, is also related directly to the sudden change of flow direction of the main channel of the Adami-Potami drainage network (Middle Pleiostocene)

- The oldest planation surfaces system, found today in altidutes greater than 240 m., was possibly created during the Oligocene - Eocene.

4. The buttes that appear on the planation surfaces (mainly on those of 160-220 m.) may be residual forms, of a now inactive conical karst. The conditions of their creation indicate a warm and humid climate, a lot different from the recent climatic conditions. Their creation took place in the lower Pliocene.

5. The buttes that appear in the southern part of the study area (Ano Sounio, Legrena, Kharakas), can be related to the 20-80m., 100-160 m. and the 160-220 m. planation surfaces, but only in a few cases.

6. Calcareous sandstones deposits are widespread, mainly in the eastern coasts of the study area (Daskalio to Sounio). They are aeolian deposits of coastal sediments of Upper Pleistocene - Lower Holocene. The fact that they can be found only in the eastern coast of Attica Peninsula reveals the possible existence of strong N-NE winds during this period, when large areas of the South Euboic gulf were terrestrial and were the supply sources for the calcareous sandstones deposits. Their diagenesis and cementation took place in brackish and marine environments at lower levels and in a terrestrial environment at the higher levels.

7. The coast line of the study area (from Kaki Thalassa to Anavyssos) ha retreated. Principally during the recent Holocene, a sea level rise of approximately 3m is observed, continuously over the last 2.500 years. This is indicated by the submarine archaeological discoveries as well as by the submarine appearances of beachrocks. The beachrocks at coastal sites are retreating and are being destroyed by the sea.

8. Human activities and their implications play a great part in forming the coastline and are considered of special interest in the study area. The Legrena, Kharakas, Sounio, Pashalimani and Pountazeza bays have had an intense tourist development. The Tourkolimano, Avlaki, Vromopigado, Daskalio and Kaki Thalassa bays are vacation residential areas with uncontrolled construction schemes. Finally, the Thorikos and Anavyssos bays, as well as the bays near Lavrio Port, are burdened by urban and industrial sewage. Lavrio Port is the main port of the area, for commercial use as well as for transportation. Human activities are not confined to the coastal area but are also present in the whole south Attica region. For example, ancient as well as the recent mining activities have alteted the natural environment.

Based on all the data mentioned above, provided that geotectonic and climatic conditions stay the same and the development in South Attica continues without a plan, the following are predicted:

- Marine transgression due to sea

level rise, already observed in the area during the Holocene.

- Sedimentation of the lower parts of the drainage networks which have occasional flow.
- Floods in certain areas after periods of heavy rainfall, due to man made structures placed perpendicularly to the flow direction of the torrents when drainage and sewers have not been provided.
- Groundwater supply wasting or quality degradation in the area, due to uncontrolled and often unreasonable use.
- Burdening of the coastal areas with additional polluting agents, from several settlements as well as from industries, which in most cases are discharged to the sea without being processed.
- Faster erosion rates in areas with high inclination that have been deforested for building purposeses.

For all these problems mentioned above, the following are proposed:

- Building construction control and land-planning.
- Installation of sewage treatment systems.
- Drainage and Sewage works.
- Monitoring and management of the area's ground water resources.
- Socioeconomic upgrade of the area after relevant studies.

Case study 5: Geomorphological study of Salar del Huasco (Chile)

The study area is located in the province of Tarapaca in northern Chile near the borders of Bolivia. The area forms a closed endorehic basin in a great altitude from 3760m to 5022m located in Altiplanos area.

Lithology is mainly constituted of volcanic rocks of the Cretaceous-Pleistocene age; alluvial, colluvium, lacustrine and aeolian deposits; and deposits of evaporites and salts. The area's main characteristic is the formation of an endorehic basin (its origin is located at the greek words: 1. ἔνδον: within and 2. ρεῖν: flow) which all together means that it is a closed water system that maintains water and does not allow it to migrate to other water systems like rivers and oceans. Usually, the water of hydrographic basins flows through the surface runoff or through the underground aquifer horizon to the sea, the ocean or other drainage networks.

On the contrary, in an endorehic basin like the one in Salar del Huasco precipitations do not outflow the basin, but they 'disappear' only through infiltration and evaporation. This process, in combination with the ultra dry climatic conditions, creates salty lakes and salt flats. At the areas of the lowest elevation in the basin salty lakes are developed and salt flats with deposits of evaporites and sulphurous, chloride, carbonate and nitric salts. Often, endorehic basins are called internal water systems.

Characteristic landforms of the Chile's relief are the Altiplanos. The Altiplanos are internal closed basins (endorehic basins) located in the central Andes in the states of Chile, Argentina, Bolivia, Peru and Equoador. They are developed in an average altitude of 3300m approximately, a slightly less than the altitude of the thibetian plateau. In comparison to the thibetian plateau, the Altiplanos are surrounded by large and active volcanoes. They extend between

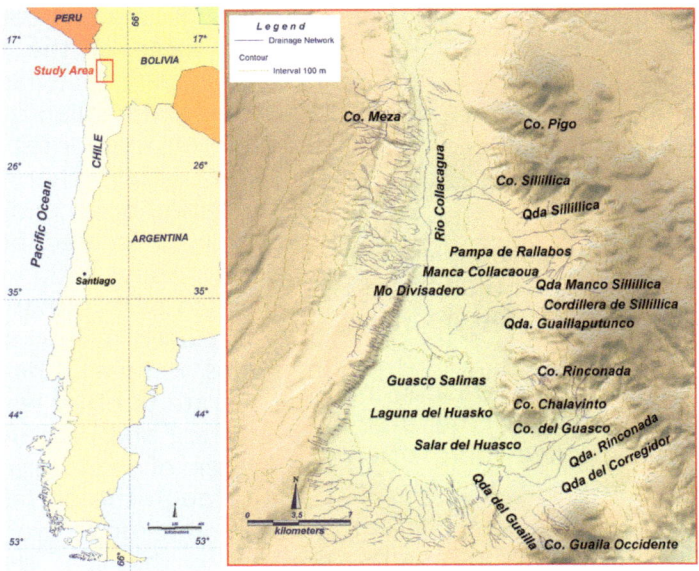

Location map

the east and west cordillera of the Andes (that form the active volcanic arc) and they are characterized by the presence of large salty lakes and fields, Quarternary deposits and volcanic rocks from the Upper Oligocene period till today. The appearances of the bedrock (Ordovisian-Creatageous age) at the surface is rare. During the end of Pleistocene, the Altiplanos area was covered by a lake. From the remaining of this lake, two more were formed, Titicaca Lake on the borders of Peru and the salty lakes Salar de Uyuni, Salar de Coiposa and Salar del Huasco in Chile .

Methodology

The geomorphological study of the area involved a series of different stages such as primary data, data analysis and creation of different thematic maps.

Regarding primary data sources, primary data were retrieved from an older geomorphological map,

the official geological map (scale 1:1.000.0000) of the National Geological Society and the Digital Elevation Model (pixel size 90m). DEM was downloaded from Shuttle Radar Topography Mission program (SRTM), which uses a Space Shuttle and obtains Earth surface data by remote sensing technology utilizing synthetic aperture radar.

Topographic and physical data (altitude, location names) and geomorphological data (drainage network, aquatic systems, slope analysis, cliffs, alluvial cones, deltaic fans, dunes, salty fields) were digitized from a pre existing geomorphological map, while geological formations and tectonic structure were digitized from the geological map.

All primary geographical, geological and environmental data were input in a Geographical Information System (GIS), where a new database was structured and updated, especially for the area of interest. Contours

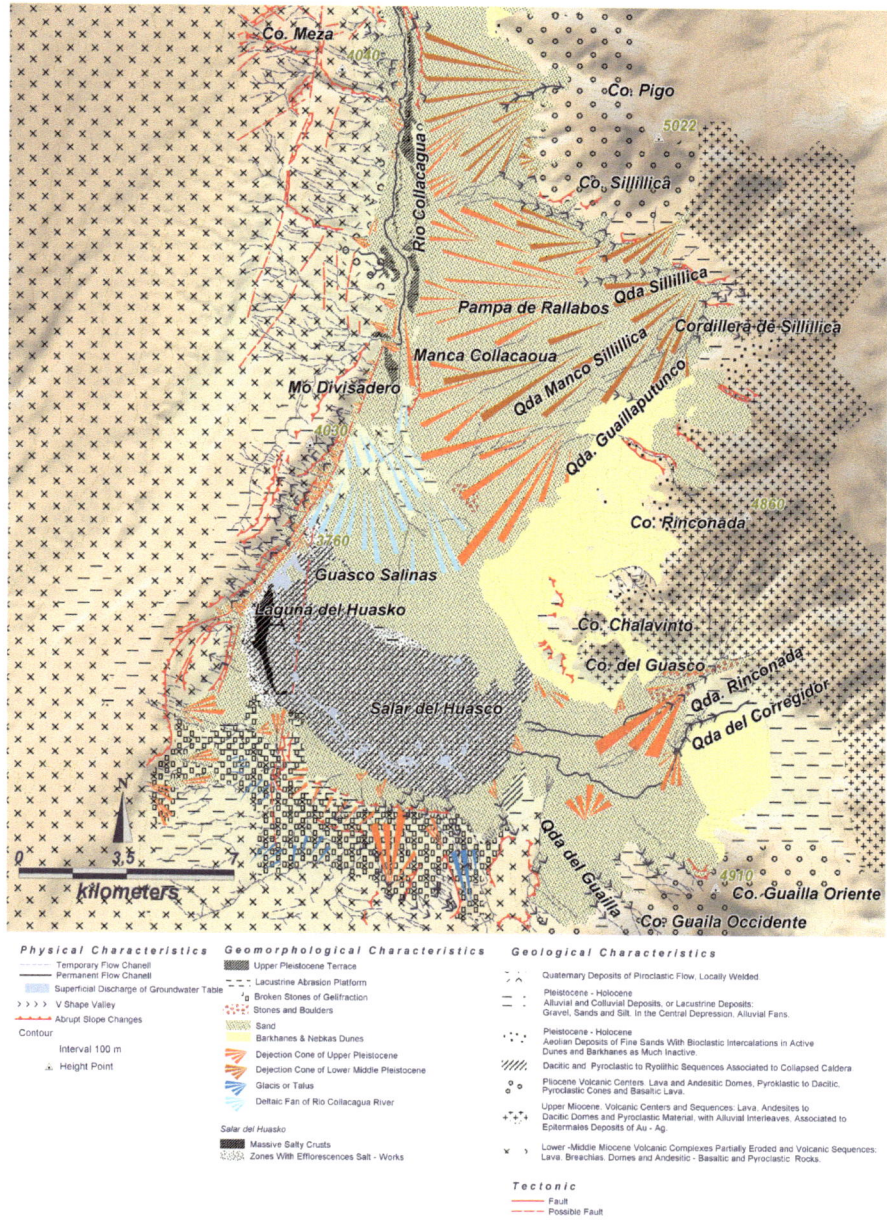

Geomorphological map of Salar del Huasco

were extracted automatically from the DEM and added to the database. After completing this step, all data were processed, statistically and spatially and the secondary data were visualized and distributed through a series of thematic maps. The study concluded into the development of the analytical geomorphological map of the area.

Climatic conditions

North Chile's climate is defined by the Anticyclon of southeast

221

Pacific, the cold ocean, current Humboldt, and the orographic affection of the Andes (e.g., Abele, 1991; Houston and Hartley, 2003). These conditions lead to ultra dry climatic conditions, in the coastal mountain range and the western steep zone, and semi dry climatic conditions, in the western mountain range. The fast rising of the Andes during Tertiary affected seriously the climatological and atmospheric conditions in central Andes. This led to an extraordinary change of the climate to being ultra drought which still affects the seasonal distribution of the rainfalls. Climate in Altiplanos is cold and dry with average annual temperatures of 3° C near the mountain range and 12° C near the lake. The average annual height of the precipitation varies between 200 mm and 600 mm. The daily temperature varies from a maximum temperature of 12° to 24° C and a minimum of -20° to +10° C. The lowest temperatures appear in the southwest areas during June-July which is winter for the South Hemisphere. Rainfall's annual cycle distributes between December and March. Maximum temperatures are also observed in this session. The rest of the year, the climate is very dry and cool, with stormy winds and sunshine. Snowfalls are observed rarely between April and December (1 to 5 occasions a year).

Ecology

Salar del Huasco is what has remained today of a 400 km Pleistocenic lake which is among others the Titicaca lake between south Peru, west Bolivia and Antofagasta region of Chile. At the time, lake Salar de Huasco was drained and its remaining looked like 'pockets of wetness in a dried sea'. The area shows a significant variety of species even in the salty fields, high endemic rate and adjustment in the local climatic conditions (rainfalls<100 mm/y, long sunshine, high sun radiation, great temperature variation). Salt lake Salar del Huasco is the only salt lake in Chile that has been used as a sanctuary by 3 threaten flamingo species of south America (Phoenicoparrus andinus, Phoeenicopterus rubber chilensis and Phoenicoparrus jamesi) 18 birds (ostrich, condor of Andes etc.) and 44 mammals (Lama, fox of Andes etc.). It also includes 203 species of endemic flora (Polypodiophyta, Pinophyta, Magnoliophyta etc.).

Geomorphology – Landforms and landscape evolution

Geology of Chile and the morphological formations are mainly a result of the orogenesis of the cordillera of Andes. An orogenesis that takes place since the subduction of Nazca plate of the east Pacific under south America's lithospheric plate began along the coastline of Chile. Besides the orogenesis, the convergence of the 2 plates, form 2 characteristic structures, the trench, which is a deep basin formed in the boundary of the 2 plates, and a series of volcanic centers (hot patches, hot spots) like the Pasha Island and the Juan Fernandez islands.

Many active volcanic centers are located in Chile and are considered to compose the limits of an area of the Pacific Ocean that is called 'ring of fire' and defines the shape of the volcanoes' spread. Some of these active volcanoes are Villarica

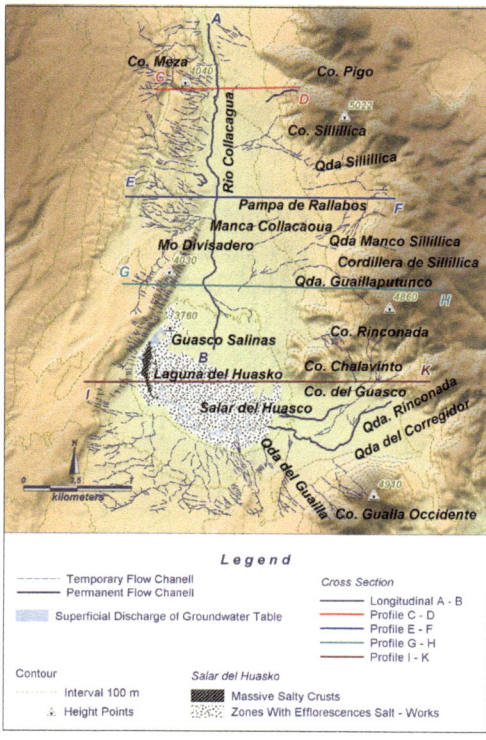

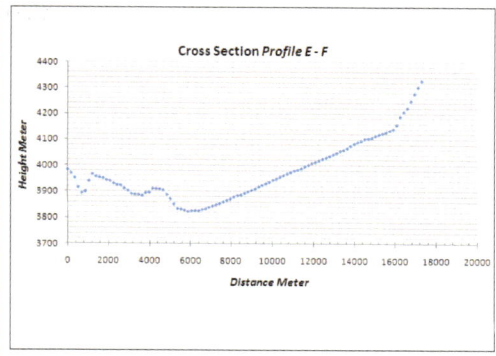

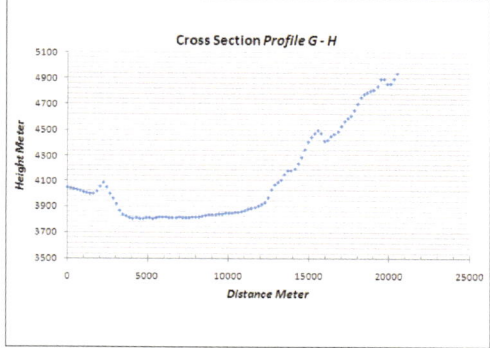

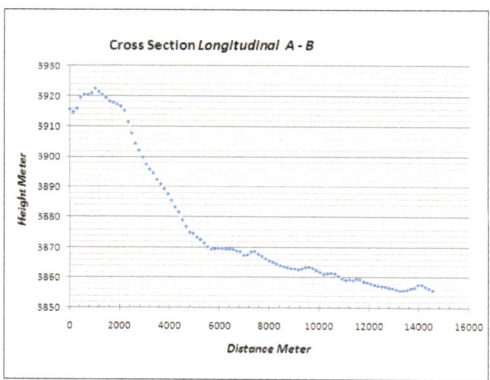

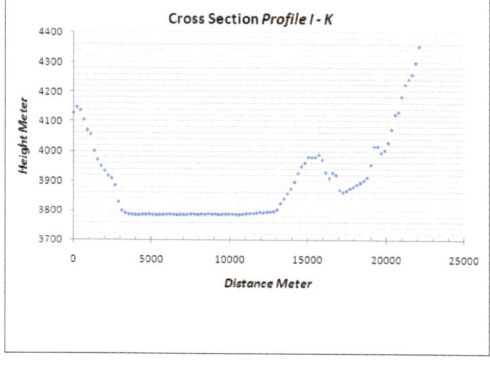

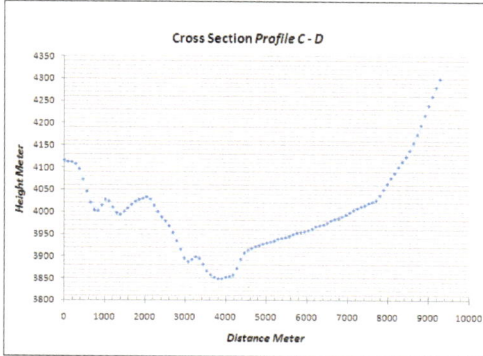

volcano and Hudson Mountain. The greatest earthquake in Chile was recorded on the 22nd of May in 1960, reaching 9.5 degrees of the Richter scale.

Climatic conditions control the spatial distribution of erosion and sediments' deposition, the development of the internal water systems (endorehic) and the sediments' 'capture' at the plateau of the Andes.

At the Altiplanos, 3 types of volcanic centers can be defined. Their spatial and temporal distribution gives significant information for the current tectonic process in the whole area. The main volcanic chain is composed by bedding volcanic complexes with great thickness andesite and dacide lavas, pyroclastic flows and volcanic domes. Earth's highest volcanic complexes can be found in this location, reaching more than 6700m of height. Great outcrops of Upper Miocene – Pleistocene volcanic rocks (andesites, dacides, ignimbrites) are located in the back – arc's plateau.

The ultra dry climate of the Neogene played significant role in controlling the evolution of the Andes' relief and preserving it. It is believed that ultra dry climate began at the end of Cenozoic. Dunai et al (2005) believes that it began somewhere between 25 and 14 My, Mortimer (1980) and Alpers and Brimhall (1988) set the ultra dry climate appearance 15-9 My and finally Hartley (2003) believes that it began much earlier, at 4-3 My. The beginning of the ultra dry period and the relative reduction of the sedimentation processes played an important role in the geodynamic evolution of the central Andes (Lamb and Davis, 2003).

Western Andes in North Chile divide in elongated geomorphological units from west to east. The coastal cordillera, the western steep zone, the western cordillera and the Altiplanos, are different geomorpohological units which are a result of the geodynamic evolution of the central and western Andes during Cenozoic, the crust's thickness and the relevant rising of the west cordillera and Altiplanos and the west-directed torsion of the western steep zone during M. Miocene (Gregory-Wodzicki, 2000).

The Altiplanos' geomorphological unit has shown a maximum structural elevation rate 0.2-0.3 mm/y from the Upper Miocene to today (Gregory-Wodzicki, 2000). The current geomorphic processes in Salar del Huasco are mechanical weathering, gravity, seasonal streams' runoff and the Aeolian and volcanic processes. The water system of Rio Collacaqua is developed between 2 cordilleras in an altitude of 4000m (Cordillera Meza) in the West and 5000m in the East (Cordilleras Piga, Sillillica, Rinconada) with a general flow direction from North to South parallel to the main fault zone of the area. The river has a fixed flow up to Manca Collacaqua and then it disappears in the deltaic deposits, supplying the underground aquifer horizon. Springs appear at the southern lower areas of the lake.

Long extended alluvial fans appear east of the central bed of the river because of the water systems in the western slopes of cordilleras Piga and Sillillica. These cordilleras are approximately 1000m higher than cordillera Meza in the West. The previously mentioned in combination with the presence of extensive alluvial fans of a 'telescopic' shape (means that older age sediments are near the mountain and getting younger as we reach the bed of the river) display structural elevation of the east mountain range during Pleistocene-Holocene. Debris and alluvial cones which can be found in the western part of the river were formed by the active tectonic of the North – South directed fault zone. The Upper Pleistocene terraces of Rio Collacaqua river were created by the change of the river's basic flow level which is the salt lake Salar del Huasco. This change was probably caused by tectonic processes and climatic changes. Scarps appear in basalts in which sometimes Mesa formations appear (Mo. Divisadero) due to the tectonic structure and the differential erosion. Fields of Barkhanes type dunes are developed northeast of the salt lake and in the northern section of the salt lake in the salty field of the pro-deltaic deposits of Rio Callacaqua.

References

Aderidan, A., Parcharidis, I., Poscolieri, M., Pavlopoulos, K., 2004: Computer-assisted discrimination of morphological units on north-central Crete (Greece) by applying multivariate statistics to local relief gradients. *Geomophology* 58, pp. 357-370.

Alpers, C.N., Brimhall, G.H., 1988. Middle Miocene climate change in the Atacama Desert, northern Chile: evidence from supergene mineralization at La Escocndida. GSA Bull. 100, 1640–1656.

Ammerman A. J., Efstratiou, N., Ntinou, M., Pavlopoulos, K., Gabrielli, R., Thomas, K. D. & Mannino, M. A., 2008: Finding the early Neolithic in AegeanThrace: the use of cores. *Antiquity*, v. 82, pp.139-150.

Athersuch, J., Horne, D.J., Whittaker, J. E., 1989: Marine and brackish water Ostracodes (Superfamilies Cypridacea and Cytheracea). Keys and notes for the identification of the species. *Synopses of the British Fauna (New series)* 43, London, 343 p.

Ayalon A., Bar-Matthews, M., & Kaufman, A., 2002: Climatic conditions during marine isotopic stage 6 in the Eastern Mediterranean region as evident from the isotopic composition of speleothems. Soreq Cave, Israel. *Geology* 30, 303-306.

Bard, E., Hamelin, B., Fairbanks, R.G. & Zindler, A. 1990: Calibration of the 14C time scale over the last 30,000 years using mass spectrometric U-Th ages from Barbados corals. *Nature* 345, 405-410.

Bar-Matthews, M., Ayalon, A. & Kaufman, A., 1997: Late Quaternary palaeoclimate in the eastern Mediterranean region from stable isotope analysis of speleothems at Soreq Cave, Israel. *Quat. Res.* 47, 155-168.

Bar-Matthews, M., Ayalon, A. & Kaufman, A., 1998: Middle to late Holocene (6500 Yr. period) paleoclimate in the Eastern Mediterranean region from stable isotopic composition of speleothems from Soreq cave, Israel. In Issar A.S. and Brown N. (eds), *Water, Environment and Society in time of climate change*. Kluer Acad. Publishers, 203-214.

Bar-Matthews, M., Ayalon, A., Gilmour, M., Matthews, A. & Hawkesworth, C.J., 2003: Sea-land oxygen isotopic relationships from planktonic foraminifera and speleothems in the Eastern Mediterranean region and their implication for paleorainfall during interglacial intervals. Geochim. *Cosmochim. Acta* 67, 3181-3199.

Bar-Matthews, M., Ayalon, A., Matthews, A., Sass, E. & Halicz, L., 1996: Carbon and oxygen isotope study of the active water-carbonate system in a karstic Mediterranean cave: implications for paleoclimate research in semiarid regions. *Geochim. . Cosmochim. Acta* 60, 337-347.

Bernier, P., Dalongeville, R., 1988: Incidence de l'activité biologique sur la cimentation des sédiments littoraux actuels. L'exemple des îles de Délos et de Rhénée (Cyclades, Grèce). *Comptes rendus de l'Académie des Sciences*, Paris, série II, 307, 1901-1907.

Bottema, S., 1982: Palynological investigations in Greece with special reference to pollen as an indicator of human activity. *Palaeohistoria* 24, 257-288.

Bottema, S., Woldring, H., 1990:

Anthropogenic indicators in the pollen record of the Eastern Mediterranean. In Bottema S., Entjes-Nieborg G. and Zeist W.V. (Eds.): *Man's role in the Shaping of the Eastern Mediterranean landscape.* A.A. Balkema, pp. 231-264.

Boyden, C. R. & Russel, P. J. C., 1972: The distribution and habitat range of the brackish water cockle (Cardium (Cerastoderma) glaucum), in the British Isles. *J. Anim. Ecol.* 43, 719-734.

Brückner, H., 1986: Man's impact on the evolution of the physical environment in the Mediterranean region in historical times. *GeoJournal* 13.1, 7-17.

Brückner, H., 1990: Changes in the Mediterranean ecosystem during antiquity – A geomorphological approach as seen in two examples. – In: Bottema, S., G.

Brückner, H., 1998: Coastal research and geoarchaeology in the Mediterranean region. - In: Kelletat, D.H. (Ed.): *German geographical coastal research – The last decade,* pp. 235-258. Institute for Scientific Co-operation, Tübingen, and Committee of the Federal Republic of Germany for the Int. Geographical Union; Tübingen.

Brückner, H., 2003: Delta evolution and culture – Aspects of geoarchaeological research in Miletos and Priene. – In: Wagner, G.A., E. Pernicka & H.P.

Bruins, H. J., MacGillivray, J.A., Synolakis, C. E., Benjamini, C., Keller, J., Kisch, H.J., Klugel, A., van der Plicht, J., 2008: Geoarchaeological tsunami deposits at Palaikastro (Crete) and the Late Minoan IA eruption of Santorini. *Journal of Archaeological Science* 35, 191-212.

Brunet, M., Desruelles, S., Cosandey, C., Fouache, E., Pavlopoulos, K., Siard, H., 2003: L'eau de Délos. *BCH (Bulletin de Correspondance Hellénique),* vol.127, pp. 516-522.

Carrion, J.S., van Geel, B., 1999: Fine-resolution Upper Weichselian and Holocene palynological record from Navarres (Valencia, Spain) and a discussion about factors of Mediterranean forest succession. *Review of Palaeobotany and Palynology* 106, 209-236.

Cayeux, L., 1911: *Description physique de Délos.* Exploration archéologique de Délos IV, Athènes-Paris, 215 p.

Cayeux, L., 1914: Les déplacements de la mer à l'époque historique. *Revue scientifique* 19, 577-586.

Chappell, J.M.A. & Polach, H.A., 1972: Some effects of partial recrystallisation on ^{14}C dating of late Pleistocene corals and mollusks. *Quat. Res.* 2, 244-252.

Church, J.A., Gregory, J.M., Huybrechts, P., Kuhn, M., Lambeck, K., Nhuan, M.T., Qin, D. & Woodworth, P.L., 2001: Changes in sea level. Chapter 11 of the Intergovernmental Panel on Climate Change Third Assessment Report, Cambridge University Press.

Clavé, B., Massé, L., Carbonel, P., Tastet, J. P., 2001: Holocene coastal changes and infilling of the La Perroche marsh (French Atlantic coast). *Oceanologica Acta* 24(4), 377-389.

Couchot, C., Fouache, E., Pavlopoulos, K., Jovanovski, M., 2007: Early Holocene environment in a subsiding balkanic graben (Skopje, FYROM): The case of tumba Madzhari (5800-

5300 BC). *Geodynamica Acta* 20/4, pp.267-274.

Cundy, A.B., Kortekaas, S., Dewez, T., Stewart, I.S., Collins, P.E.F., Croudace, I.W., Maroukian, H., Papanastassiou, D., Gaki-Papanastassiou, P., Pavlopoulos, K. & Dawson, A., 2000: Coastal wetlands as recorders of earthquake subsidence in the Aegean: a case study of the 1894 Gulf of Atalanti earthquakes, central Greece. *Mar. Geol.* 170, 3-26.

Dalongeville R. & Sanlaville P., 1984: *Essai de synthèse sur le beachrock.* In Dalongeville R. (ed), *Le beachrock: actes du colloque,* Lyon 1983, T.M.O. 8, Lyon, Maison de l'Orient Méditerranéen, 161-167.

Dalongeville, R., Desruelles, S., Fouache, E., Hasenohr, C., Pavlopoulos, K., 2007: Hausse relative du niveau marin a Delos (Cyclades, Grece): rythme et effets sur les paysages littoraux de la ville helenistique. *Mediterranee*, vol 108, pp. 17-28.

Desruelles, S., Fouache, E., Dalongeville, R., Pavlopoulos, K., Peulvast, J. P., Coquinot, Y., Potdevin, J. L., Hasenohr, C., Brunet, M., Mathieu, R., Nicot, E., 2007: Sea-level changes and shoreline reconstruction in the ancient city of Delos (Cyclades, Greece). *Geodynamica Acta* 20/4, pp.231-239.

Desruelles, S., Fouache, E., Pavlopoulos, K., Dalongeville, R. Peulvast, J.P., Coquinot, Y., Potdevin, J.-L., (in press) Recent sea-level change and beachrock in the: the insular group of Mykonos–Delos–Rhenia (Cyclades, Greece). *Géomorphologie.*

Desruelles, S., Fouache, E.,

Pavlopoulos, K., Dalongeville, R., Peulvast, J.P., Coquinot, Y., Potdevin, J.L., 2004: Beachrocks et variations récentes de la ligne de rivage en Mer Égée dans l'ensemble Mykonos-Délos-Rhénée (Cyclades, Grèce). *Géomorphologie: relief, processus, environnement* 1, 5-18.

Desruelles, S., Fouache, E., Pavlopoulos, K., Dalongeville, R., Peulvast, J. P., Coquinot, Y. et Potdevin, J. L., 2004: Beachrocks et variations recentes de la ligne de rivage en Mer Egee dans l ensemble insulaire Mykonos-Delos-Rhenee (Cyclades, Grece). *Geomorphologie* 1, p.5-18.

Dominey-Howes, D., Dawson, A. & Smith, D., 1998: Late Holocene coastal tectonics at Falasarna, western Crete : a sedimentary study, in *"Coastal tectonics"*, Geological Society, Londres, Stewart I. and Vita-Finzi Cl. eds, 146, pp. 343-352.

Dominey-Howes, D.T.M., Cundy A.B. & Croudace, I.W., 2000: High Energy Marine Flood Deposits on Astypalaea Island, Greece: Possible Evidence for the AD1956 southern Aegean Tsunami. *Mar. Geol.* 163, 303-315.

Dufaure, J.J. and Zamanis, A., 1980: Styles néotectoniques et étagements de niveaux marins sur un segment d'arc insulaire, le Péloponnèse. Proc. Actes Coll. CNRS '*Niveaux marins et tectonique quaternaire dans l' aire méditerranéenne"*, 77-107.

Dunai, T.J., Gonzales Lopez, G.A., Juez-Larre, J., 2005. Oligocene–Miocene age of aridity in the Atacama Desert revealed by exposure dating of erosion-sensitive landforms. Geology 33 (4), 321–324.

Fairbanks, R. G., 1989: A 17.000-year glacio-eustatic sea level record: influence of glacial melting rates on the Younger Dryas event and deep-ocean circulation, *Nature* 342, 637-642.

Fairbridge, R.W., 1961: Eustatic changes in sea level. *Physics & Chemistry of the Earth* 4, 99-185.

Feldskaar, W. & Cathles, L., 1991: The present rate of uplift of Fennoscandia implies a low-voscosity asthenosphere. *Terra Nova* 3, 393-400.

Fleming, K., Johnston, P., Zwartz, D., Yokohama, Y. Lambeck, K. & Chappell, J., 1998: Refining the eustatic sea level curve since the Last Glacial Maximum using far and intermediate field sites, *Earth and Planetary Science Letters* 163, 327-342.

Flemming N.C., 1979: Archaeological indicators of sea-level. In Les indicateurs de niveaux marins, séminaire du 2 décembre 1978. *Océanis. Fascicules Hors-Série*, volume 5, 184-191.

Flemming, N.C., 1972: Eustatic and tectonic factors in the relative vertical displacement of the Aegean coast. – In: Stanley, D.J. (ed.): *The Mediterranean Sea*. – Stroudsberg/ Pennsylvania, Dowden, Hutchinson & Ross: 189-201.

Flemming, N.C., Czartoryska, N.M. and Hunter, P.M., 1973: Archaeological evidence for eustatic and tectonic components of relative sea level sea level change in the South Aegean. *Colston Papers* 23, 1-66.

Fouache, E., Desruelles, S., Pavlopoulos, K., Dalongeville, R., Peulvast, J.P., Potdevin, J.L., 2005: Using beachrocks as sea level indicators in the insular group of Mykonos, Delos and Rhenia (Cyclades, Greece). In Fouache E. and Pavlopoulos K. (Ed.) *Z.Geomorphoph. N. F. Suppl.* Vol.137 pp.37-43.

Garcia, M., Herail, G., 2005. Fault-related folding, drainage network F. Kober, S. Ivy-Ochs, F. Schlunegger, H. Baur, P.W. Kubik , R. Wieler, 2007. "Denudation rates and a topography-driven rainfall threshold in northern Chile: Multiple cosmogenic nuclide data and sediment yield budgets". Geomophology, 83, pp. 97-120.

Godwin, H., 1962: Half-life of radiocarbon. *Nature* 195, 984.

Gregory-Wodzicki, K.M., 2000. Uplift history of the Central and Northern Andes: a review. GSA Bull. 112, 1091–1105.

Hartley, A.J., 2003. Andean uplift and climate change. J. Geol. Soc. (Lond.) 160, 7–10.

Hejl, E., Riedl, H., Weintgartner, H., 2002: Post-plutonic unroofing and morphogenesis of the Attic-Cycladic complex (Aegea, Greece). *Tectonophysics* 349, 37-56.

Hopley, D., 1986: Beachrock as sea-level indicator. – In: Van der Plassche, O. (ed.): *Sea-Level Research: a manual for the collection and evaluation of data*. Amsterdam, 157-173.

Kalpaxis, Th., Athanassas, K., Bassiakos, I., Brennan, T., Hayden, B., Nodarou, E., Pavlopoulos, K. & Sarris, A., 2006: Preliminary results of the Istron, Mirabello, geophysical and geoarchaeological project, 2002-2004. *The Annual of the British School at Athens*, vol 101, pp. 131-181.

Kanellopoulos, Th., Kapsimalis, V., Poulos, S., Angelidis, M., Karageorgis, A., Pavlopoulos, K., 2008: The influence of the Evros River on the recent sedimentation of the inner shelf of the NE Aegean Sea. *Environmental Geology* 53, pp. 1455-1464.

Katsikatsos, G., 1992: *Geology of Greece*. Athens.

Kayan, I., 1988: Late Holocene sea level changes on the western Anatolian coast. *Paleogeography, Paleoclimate and Paleoecology* 68, 205-218.

Kershaw, S. and Guo, Li., 2001: Marine notches in coastal cliffs: indicators of relative sea-level change, Perachora Peninsula, central Greece. *Marine Geology* 179, 213-228.

Kissel, C., Laj, C, Mazaud, A., 1986a: First paleomagnetic results from Neogene formations in Evia, Skyros and the deformation of central Aegean. *Geophysical Research Letters* 13(13), 1446-1449.

Klige, R.K., 1980: The Level of the Ocean in the Geological Past. Nauka, Moscow, 111 pp.

Kraft, J.C., Kayan I., Brückner, H. & Rapp, G., 2000: A geologic analysis of ancient landscapes and the harbors of Ephesus and the Artemision in Anatolia. *Jahreshefte des Österreichischen Archäologischen Institutes* 69, 175-233, Vienna.

Kraft, J.C., Kayan, I. & Brückner, H., 2001: The geological and paleogeographical environs of the Artemision. – In: Muss, U. (ed.): *Der Kosmos der Artemis von Ephesos. Österreichisches Archäologisches Institut, Sonderschriften*, Bd. 37: 123-133; Vienna.

Kraft, J.C., Kayan, I., Brückner, H. & Rapp, G., 2003: Sedimentary facies patterns and the interpretation of paleogeographies of ancient Troia. – In: Wagner, G.A.,

Laborel, J., Morhange, C., Lafont, R., Le Campion, J., Laborel-Deguen, F. & Sartoretto, S., 1994: Biological evidence of sea level rise during the last 4500 years, on the rocky coasts of continental southwestern France and Corsica, *Marine Geology* 120, pp. 203-223.

Lamb, S., Davis, P., 2003. Cenozoic climate change as possible cause of the rise of the Andes. Nature 425, 792–797.

Lambeck, K. & Purcell, A., 2005: Sea level change in the Mediterranean Sea, since the LGM: model predictions for tectonically stable areas. *Quat. Sci. Rev.* 24, 1969-1988.

Lambeck, K. and Bard, E., 2000: Sea level changes along the French Mediterranean coast for the past 30000 years. *Earth and Planetary Science Letters* 175, pp. 203-222.

Lambeck, K., 1996: Sea-level changes and shore-line evolution in Aegean Greece since Upper Paleolithic time. *Antiquity* 70, 588-610.

Lambeck, K., 1996: Sea-level changes and shoreline evolution in Aegean Greece since Upper Palaeolithic time. *Antiquity* 70 269, 588-611.

Lambeck, K., Johnston, P., 1995: Land subsidence and sea-level change: contributions from the melting of the last great ice sheets and the isostatic adjustement of the Earth. In Barends F. J., Brower F. J. J., Schroder F. H. (Eds): *Land Subsidence, Proc. Fifth Int. Symp.*

on Land Subsidence, The Hague, 16-20 Oct 1995. Balkema, Rotterdam, 3-18.

Lauritzen S. E. & Lundberg, J., 1999: Calibration of the speleothem data function: an absolute temperature record for the Holocene in northern Norway. *Holocene* 9, 659-669.

Le Beach-Rock, 1984: Colloque, Lyon, Novembre 1983. Travaux de la Maison de lÓrient 8, Lyon, 197 pp.

Le Pichon, X. and Angelier, J., 1979: The Hellenic Arc and Trench system: A key to the neotectonic evolution of the Eastern Mediterranean area. *Tectonophysics* 60, 1-42.

Lerman, J.C., 1972: Carbon-14 dating: origin and correction of isotope fractionation errors in terrestrial living matter. *Proc. 8th Inter. Conf. on Radiocarbon Dating*, 2. Royal Soc. Of New Zealand, Wellington, 613-624.

Libby, W.F., 1955: Radiocarbon Dating. Chicago, University of Chicago Press.

Maroukian, H., Gaki-Papanastassiou, K., Karymbalis, E., Vouvalidis, K., Pavlopoulos, K., Papanastassiou, D., Albanakis, K., 2008: Morphotectonic control on drainage network evolution in the Perachora Peninsula, Greece. *Geomorphology*, In press 2008, pp.1-12.

Maroukian, H., Gaki-Papanastassiou, K., Pavlopoulos, K., Sabot, V., 2004: The assumed future sea level rise as a natural hazard threatenig the coastlines of continental Greece. *Annales Geologiques des Pays Helleniques*, v.40, pp. 69-82.

Mazzini, I., Anadon, P., Barbieri, M., Castorina, F., Ferreli, L., Gliozzi, E., Mola, M. & Vittori, E., 1999: Late Quaternary sea-level changes along the Tyrrhenian coast near Orbetello (Tuscany, central Italy): palaeoenvironmental reconstruction using ostracodes. *Marine Micropaaeondology* 37(3-4), 289-311.

Mercier, J., Delibassis, N., Gauthier, A., Jarrige, J., Lemeille, F., Philip, H., Sebrier, M. and Sorel, D., 1979 : La néotectoniquede l' arc Egéen. *Rev. Géol. Dyn. Et Géogr. Phys.* 21, 67-92.

Montgomery, D.R., Balco, G., Willett, S.D., 2001. Climate, tectonics,and the morphology of the Andes. Geology 29, 579–582.

Morhange, C., Laborel, J. & Hesnard, A., 2001: Changes of relative sea level during the past 5000 years in the ancient harbour of Marseilles, Southern France, *Palaeogeography, palaeoclimatology, palaeooecology* 166, pp. 319-329.

Mörner, N. A., 1976b: Eustatic changes during the last 8,000 years in view of radiocarbon calibration and new information from the Kattegatt region and other northwestern European coastal areas. Palaeogeogr. *Palaeoclim. Palaeoecol.* 19, 123-151.

Mörner, N. A., 1996: Rapid changes in coastal sea level. *J. Coastal Res.* 12, 797-800.

Mörner, N. A., 2000: Sea level changes and coastal dynamics in the Indian Ocean. Integrated Coastal Zone Management, *Launch Issue*, 17-20.

Mörner, N. A., 2003: Sea level changes in the Past, at Present and in the Near-Future. Global aspects. Observations versus Models. *GI2S Coast, Research Publication* 4, 5-9. (IGCP-437).

Mörner, N. A., 2004: Estimating future sea level changes. Global Planet. *Change* 40, 49-54.

Mörner, N.-A., 1996: Sea level variability, *Z. Geomorh. N. F.* 102, pp. 223-232.

Mortimer, C., 1973. Cenozoic history of the southern Atacama Desert,Chile. J. Geol. Soc. (Lond.) 129, 505–526.

Mortimer, C., 1980. Drainage evolution in the Atacama Desert of northernmost Chile. Rev. Geol. Chile 11, 3–28.

Mortimer, C., Saric, N., 1975. Cenozoic studies in northernmost Chile.Geol. Rundsch. 64, 395–420.

Mortimer, C., Farrar, E., Saric, N., 1974. K–Ar ages from Tertiary lavas of the northernmost Chilean Andes. Geol. Rundsch. 63,484–490.

Negris, P. 1903: *Régression et transgression de la mer depuis l'époque glaciaire jusqu'à nos jours.* Revue Universitaire des Mines, 3, 249-281.

Negris, P., 1904a: *Vestiges archéologiques submergés.* Mitteilungen des Deutschen Archäologischen Instituts, 29, 340-363.

Negris, P., 1904b: Nouvelles observations sur la dernière transgression de la Méditerranée. *C. R. Acad. Sci., II*, 379-381.

Neumeier, U., 1998: *Le rôle de l'activité microbienne lors de la cimentation précoce des beachrocks (dépôts marins littoraux).* Thèse de Doctorat, Université de Genève, Suisse, 152 p.

Palyvos, N., Pantosti, D., DeMartini, P. M., Lemeille, F., Sorel, D., Pavlopoulos, K, 2005: The Aigion-Neos Erineos normal fault system (Western Corinth Gulf Rift, Greece): Geomorphological signature, recent earthquake history and induced coastal changes during the Holocene. *Journal of Geophysical Research* 110, pp.1-15.

Papanikolaou, D., 1986: *Geology of Greece*, Athens.

Papanikolaou, D., Lykousis, V., Chronis, G. and Pavlakis, P., 1988: A comparative study of neotectonic basins across the Hellenic arc: the Messiniakos, Argolicos, Saronikos and Southern Evoicos gulfs. *Basin Research* 1, 167-176.

Papazachos, B.C., Koutitas, Ch., Hatzidimitrou, P.M., Karacostas, B.G. & Papaioannou, Ch., A., 1986: Tsunami hazard in Greece and the surrounding area. *Annal. Geophys.* 4, B(1), 79-90.

Pavlopoulos, K. Karkanas, P., Triantaphyllou, M., Karymbalis, E., Tsourou, Th. & Palyvos, N., 2006: Palaeoenvironmental evolution of the coastal plain of Marathon, Greece, during the Late Holocene: Deposition environment, climate and sea-level changes. *Journal of Coastal Research,* v. 22, n.2, pp. 424-438.

Pavlopoulos, K., Karkanas, P., Triantaphylloy, M. & Karymbalis, E., 2003: Climate and sea level changes recorded in coastal plain of Marathon, Greece. In Fouache (Ed.), The Mediterranean World Environment and History. Elsevier Paris, 453-465.

Pavlopoulos,K.,Maroukian,H.,1998: Geomorphic and morphotectonic Observations in the drainage network of Kakotopia Stream, North East Attica, Greece. *Geologica Balcanika,* vol. 27, no 3-4, pp.55-60.

Pavlopoulos, K., Maroukian, H., Zamani, A, 1993: Coastal Retreat in the Plain of Marathon (East Attica) Greece: Cause and Effects. *Geologica Balcanika* 23(2), 67-71.

Pavlopoulos, K., Theodorakopoulou, K., Bassiakos, I., Hayden, B., Tsourou, Th., Triantaphyllou, M., Kouli, K., Vandarakis, D., 2007: Paleoenvironmental evolution of Istron (N.E Crete), during the last 6000 years: depositional environment, climate and sea level changes. *Geodynamica Acta* 20/4, pp.219-229.

Pavlopoulos, K., Triantafyllou, M., Karymbalis, E., Karkanas, P., Kouli, K., Tsourou, Th., 2007: Landscape evolution recorded in the embayment of Palamari (Skyros Island, Greece) from the beginning of the Bronze Age until recent times. *Geomorphologie* vol. 1/2007, pp. 37-48.

Pavlopoulos, K., Triantaphyllou, M., Karymbalis, E., Karkanas, P., Kouli, K. & Tsourou, Th., 2007: Landscape evolution recorded in the embayment of Palamari (Skyros Island, Greece) from the beginning of the Bronze Age until recent times. *Geomorphologie* no1, 2007, pp.37-48.

Pedersen, G., Gjevik, B., Harbitz, C.B., Dybesland, E., Johnsgard, H. & Langtangen, H.P., 1995: GITEC Final Scientific Report (Chapter Nine). In: Tinti, S. (Ed.) *The Genesis and Impact of Tsunamis on the European Coasts*, Final Scientific Report.

Peltier, W. R., 1998: Postglacial variations in the level of the sea: implications for climate dynamics and soild-earth geophysics. *Reviews of Geophysics* 36, 603-689.

Pirazzoli, P. A., 1991: World atlas of Holocene sea level changes, Elsevier, *Oceanography series* 58, 1-300.

Pirazzoli, P. A., Ausseil-badie, J., Giresse, P., Hadjidaki, E. & Arnold, M., 1992: Historical environmental changes at Phalasarna harbor, West Crete, Geoarcheology, *An International Journal* 7, pp. 371-392.

Pirazzoli, P. A., Montaggioni, L. F., Salvat, B. & Faure, G., 1988: Late Holocene sea level indicators from twelve atolls in the central and eastern Tuamotus (Pacific Ocean). *Coaral Reefs* 7, 57-68.

Pirazzoli, P. A., Stiros, S. C., Arnold, M., Laborel, J., Laborel-Deguen, F. & Papageorgiou, S., 1994: Episodic uplift deduced from Holocene shorelines in the Perachora Peninsula, Corinth area, Greece. *Tectonophysics* 229, 201-209.

Pirazzoli, P. A., Thommeret, J., 1973: Une donnée nouvelle sur le niveau marin à Marseille à l'époque romaine, *Comptes Rendus de l'Académie des Sciences de Paris* 227, pp. 2125-2128.

Pirazzoli, P.A., 1986: The Early Byzantine Tectonic Paroxysm. *Zeitschrift für Geomorphologie, NF, Suppl.Bd.* 62, 31-49.

Pirazzoli, P.A., 1991: World Atlas of Holocene sea-level changes. Elsevier *Oceanography Series* 58, Amsterdam.

Pirazzoli, P.A., Laborel, J. & Stiros, S.C., 1996a: Coastal indicators of rapid uplift and subsidence: examples from Crete and other eastern Mediterranean sites. Zeits. *Geomorphol., Suppl.* 102, 21-35.

Pirazzoli, P.A., Laborel, J. & Stiros, S.C., 1996b: Earthquake clustering in the Eastern Mediterranean during historical times. *J. Geophys. Res.*

101, B3: 6083-6097.

Pirazzoli, P.A., Stiros, S.C., Arnold, M., Laborel, J. & Laborel-Deguen, F., 1999: Late Holocene coseismic vertical displacements and tsunami deposits near Kynos, Gulf of Euboea, Central Greece. *Phys. Chem. Earth (A)* 24, 361-367.

Pirazzoli, P.A., Stiros, S.C., Arnold, M., Laborel., J., Laborel-Deguen, F., and Papageorgiou, S., 1994: Episodic uplift deduced from Holocene shorelines in the Perachora Peninsula, Corinth area, Greece. *Tectonophysics* 229, 201-209.

Schilman B., Bar-Matthews, M., Almogi-Labin, A. & Luz, B., 2001b: Global climate instability reflected by Eastern Mediterranean marine records during the late Holocene, *Palaeogeogerap. Palaeoclimatol. Palaoecol.* 176, 157-176.

Shackleton, N.J., Duplessy, J-C., Arnold, M., Maurice, P., Hall, M.A. & Cartlidge, J., 1988: Radiocarbon age of Last Glacial Pacific deep water. *Nature* 335, 708-711.

Sokac, A., 1978: Pleistocene ostracode fauna of the Pannonian Basin in Croatia. *Palaeontology Jugoslavia* 20, 1-51.

Stewart, I. S. & Vita-Finzi, C., 1996: Coastal uplift on active normal faults: the Eliki Fault, Greece. *Geophys. Res. Lett.* 23, 1853-1856.

Stiros, S., 1986: Geodetically controlled taphrogenesis in back-arc environments: three examples from central and northern Greece. *Tectonophysics* 130, 281-288.

Stiros, S., 1988: Model for the N. Peloponnesian (C. Greece) uplift. *J. Geodyn.* 9, 199-214.

Stiros, S., Pirazzoli, P., Rothaus, R., Papageorgiou, S., Laborel, J. and Arnold, M., 1996: On the date of construction of Lechaeon, western harbor of ancient Corinth, Greece. *Geoarchaeology* 11, 3, 251-263.

Stiros, S.C. & Pirazzoli, P.A., 1998: Late Quaternary coastal changes in the Gulf of Corinth, Greece: tectonics, earthquakes, archaeology. Geodesy Laboratory, Dept. Civil Engineering, Patras University, 49 p.

Stiros, S.C. and Pirazzoli, P.A., 1998: *Late Quaternary coastal changes in the Gulf of Corinth Greece. Tectonics, earthquake, archaeology. Gulf of Corinth.* Field Trip Guide Book, 48pp.

Stiros, S.C., Arnold, M., Pirazzoli, P.A., Laborel, J., Laborel, F. & Papageorgiou S., 1992: Historical coseismic uplift on Euboea Island, Greece. *Earth Planet. Sci. Lett.* 108, 109-117.

Stiros, S.C., Laborel, J., Laborel-Deguen, F., Papageorgiou, S., Evin, J. & Pirazzoli P.A., 2000: Seismic coastal uplift in a region of subsidence: Holocene raised shorelines of Samos Island, Aegean Sea, Greece. *Marine Geology* 170, 41-58.

Stiros, S.C., Marangou, L. & Arnold, M., 1994: Quaternary uplift and tilting of Amorgos Island (southern Aegean) and the 1956 earthquake. *Earth Planet. Sci. Lett.* 128, 65-76.

Stuiver R. M., Reimer, P. J., Bard, E., Beck, J. W., Burr, G. S., Hughen, K. A., Kromer, B., Mc Cormac, G., Van Der Plicht, J., Spurk, M., 1998: INTCAL98 radiocarbon age calibration, 24,000-0 cal BP. *Radiocarbon* 40, 3, 1041-1083.

Theodorakopoulou, K., Pavlopoulos, K., Tsourou, Th., Triantaphyllou, M., Kouli, K., Vandarakis, D., Bassiakos,

Y., Hyden, B., 2005: Coastal changes and human activities at Istron-Kalo Chorio (NE Crete, Greece) during the Upper Holocene. *Revista de Geomorfologie* v.7, pp. 21-31.

Thommeret, Y., Thommeret, J., Laborel, J., Montaggioni, L.F. & Pirazzoli, P.A., 1981: Late Holocene shoreline changes and seismo-tectonic displacements in western Crete (Greece). *Zeits. Geomorphol., Suppl.* 40, 127-149.

Triantaphyllou M.V., Gogou A., Lykousis V., Bouloubassi I., Ziveri P., Emeis K.-C., Rosell-Melé A., Kouli K., Dimiza M., Katsouras G., Nunez N.G., Papanikolaou M., Dermitzakis M.D. (submitted) Holocene primary production archive and landborne imprints: micropaleontological and geochemical evidence for sapropel formation and climate variability in the SE Aegean Sea. *Palaeogeogr. Palaeoclimatol. Palaeoecol.*

Triantaphylloy, M., Pavlopoulos, K., Tsourou, Th. & Dermitzakis, M., 2003: Brackish marsh benthic microfauna and paleoenvironmental changes during the last 6000 years on the coastal plain of Marathon (SE Greece). *Rivista Italiana di Paleontologia e Stratigrafia*, vol. 109, n. 3, pp. 539-547.

Tsimplis, M.N. & Baker, T.F., 2000: Sea level drop in the Mediterranean Sea: An indicator of deep water salinity and temperature changes? *Geophys. Res. Let.* 27 (12), 1731-1734.

Tsimplis, M.N. & Blackman, D., 1997: Extreme sea-level distribution and return periods in the Aegean and Ionian Seas. *Est. Coast. Shelf Sci.* 44, 79-89.

Tsimplis, M.N. & Josey, S.A.: 2001 Forcing of the Mediterranean Sea by atmospheric oscillations over the North Atlantic. *Geophys. Res. Let.* 28(5), 803-806.